Advance Praise for Ubernomics

"*Ubernomics* is über-needed by any organization that wants to be around in the next five years. Read it and you'll be here for fifty—and more! Barbara Gray is half brilliant analyst and half seer. The result is a book that is both crystal clear and a crystal ball."
—Joey Reiman, Chairman, BrightHouse and Managing Director, The Boston Consulting Group

"Barbara is one of the most astute and forward-looking analysts who covered lululemon. *Ubernomics* gave me a framework to think about the sharing economy capital structure and the value that can be gained from that."
—Christine Day, CEO, Luvo (former CEO, lululemon)

"Barbara's overall analysis centering on the three new values of advocacy, connection and collaboration is very powerful. And the examples, both of firms born in the new economy and others trying to adapt to it, are fascinating."
—Jean-Claude Larreche, Professor of Marketing, INSEAD, and author of *The Momentum Effect*

"Barbara has an impressive vision for the future. *Ubernomics* is fast-paced and a great foreshadowing of how technology and the empowered consumer will re-shape the Fortune 1000."
—Andrew Reid, Founder, Vision Critical

"Thoroughly researched, *Ubernomics* highlights the trends that are rapidly up-ending the ways customers, companies, and communities are interacting and building new business models. To ignore, defer or arrogantly dismiss understanding this shift towards ubernomics will come at a price; just ask Blockbuster, the taxi industry and a slew of other companies hiding behind regulatory or monopolistic practices of the past. I'm looking forward to the next book as a retrospective in three years!"
—Adam Broadway, Founder/CEO, Near Me

"When I helped start Thumbtack many years ago, I never imagined that we were on the cutting edge of an economic movement far larger than we knew. Barbara Gray describes this movement towards economic abundance in great detail, with stories and case studies that will resonate with any reader."
—Sander Daniels, Co-Founder, Thumbtack

"*Ubernomics* is a master class on the real economy for would-be disrupters and industry incumbents alike."
—Mark Gilbreath, CEO/Founder/Skipper, LiquidSpace

"Barbara's intense interest and proven intuition into the new breed of businesses who are activating the newfound reality of abundant supply/demand makes *Ubernomics* incredibly convincing, fun and challenging all in one."
—Cameron Doody, Co-Founder, Bellhops

Barbara Gray

Ubernomics

How to Create Economic Abundance and Rise above the Competition

Oct 26, 2018

Tom,

Enjoy Ubernomics!

Barb

First published by Barbara Gray in 2016

Copyright © Barbara Gray, 2016

All rights reserved. No part of this publication may be reproduced, stored, or transmitted in any form or by any means, electronic, mechanical, photocopying, recording, scanning, or otherwise without written permission from the publisher. It is illegal to copy this book, post it to a website, or distribute it by any other means without permission.

First Edition

ISBN: 978-0-9953420-0-2

This book was professionally typeset on Reedsy
Find out more at reedsy.com

To my husband, Greg, and my boys, Brady and Adam, who fill my life with abundance.

Contents

Introduction: An Analyst's Journey

When I decided to leave the corporate world in 2010, I was the proud owner of a BlackBerry, I had never tweeted or blogged, and I dressed in Armani suits. I really had no strategic plan or idea of where my research would take me, but I was driven by a strong intellectual curiosity and a passion for reading business strategy books. I launched Brady Capital Research with the lofty aspiration to create an investment research firm "for the next generation of investors"(as my company tagline reads) by focusing on innovative and high-growth companies that are making a positive difference in the world.[1] At the time, I had no clue that I was embarking on a six-year journey of pioneering research into the emerging field of abundance economics (i.e., ubernomics) and that all the information and knowledge I amassed would lead to the creation of this book.

At the start of 2016, I decided to reread the 30-plus research reports and articles I had written during the previous six years, realizing they basically tracked the evolution of my discovery of my ubernomics theory, which forms the basis of this book. When I shared this realization with my husband, Greg, he recommended I take a look at Joseph Campbell's *The Hero's Journey.* As I started to research the different stages and archetypes Campbell describes in his book, I was amazed to find that my journey of discovery shared many themes.

When I left the luxury and comfort of what Campbell would call my Ordinary World to explore the Underworld of Capitalism,

I encountered individuals from the "tribes" of Social Responsible Investing, Conscious Capitalism, Impact Investing, and the Sharing Economy. These tribespeople held very different ideologies and beliefs than did the investment professionals and corporate executives I was used to working with, and they provided me with deep insights for understanding the new social value drivers.

The most influential players I encountered were the authors of the some 200 business strategy books I had read during that time. In terms of Campbell's archetypes, they served as mentors by guiding me along the way with their knowledge and insights. And the players I continue to admire are the founders of companies like lululemon, Starbucks, Whole Foods Market, Chipotle Mexican Grill, LinkedIn, Zillow, Airbnb, Uber, and the other sharing economy companies that are creating movements rather than just marketplaces. These can be considered the tricksters, as they relish the disruption of the status quo and turn the Ordinary World into chaos by defying traditional economic principles of scarcity. I am so appreciative of the knowledge I gained from all the players I encountered on my journey, and you will see throughout the book how their different insights and ideas influenced my thinking and the conclusions I've drawn on the emerging field of abundance economics.

In Part I, I introduce the concept of ubernomics via a dot-com déjà vu, talk about how this new collaborative form of e-commerce is a wake-up call for CEOs, and discuss how companies like Airbnb and Uber are defying traditional economic principles of scarcity. In Part II, I explore how the democratization of influence, data, and physical and human capital has led to a structural shift in societal, technological, and economic forces, leading to the emergence of the three new social value drivers of advocacy, connection, and collaboration. And in Part III, I look at how companies can achieve ubernomics through creating long tails of supply and blue oceans of demand, and discuss how ubernomics is the next generation of business strategy.

Readers of this book can be assured that, given my background as a former top-ranked sell-side equity analyst and fueled by my profound curiosity and passion for coming up with new ways to look at and value companies, I have based my conclusions on thorough and comprehensive research. By sharing my discovery of the three new social value drivers that are enabling companies to defy traditional economic principles of scarcity, it is my hope that readers will gain a better understanding and appreciation of ubernomics and its implications—implications which, if you hope to maximize your business success rather than fall behind in this new economic reality, you must recognize and embrace.

I

Introduction to Ubernomics

1

The Next Generation of E-Commerce

Dot-Com Déjà Vu

At drinks with colleagues after work one evening, I was talking with the bartender, who sketched out on a napkin for me the concept of the company he was planning to start and take public. But he wasn't looking to launch the next "Airbnb of x" or "Uber of y," for this was back in early March 2000, just days before the burst of the dot-com bubble.

Whereas dot-com companies were based on the first generation of e-commerce, a one-to-many business model, marketplace companies like Airbnb and Uber are based on the next generation of e-commerce, a many-to-many business model. I am becoming increasingly skeptical that the marketplace model, or e-commerce 2.0, will translate for most sharing/on-demand economy start-ups, just as the e-commerce model did not translate for most of the dot-com start-ups.

We are starting to see a perfect storm in start-up land as minimal barriers to entry and a recent flood of venture capital lead to increased competition from existing players and new entrants. Applying Porter's Five Forces (a competitive framework developed by Michael E. Porter of Harvard Business School that assesses the threat of competition from substitutes, established rivals, and new entrants, as well as the bargaining power of suppliers and customers), I see this is creating a double jeopardy for marketplace start-ups, as it is leading to a simultaneous increase in bargaining power for both suppliers and buyers, and resulting in rising acquisition costs and attrition on both the supply and demand sides. This threatens to expose the weak economic fundamentals of many start-ups that have relied on venture capital funding to subsidize their costs of acquiring both suppliers and customers.

Although most dot-com start-ups met the fate of Pets.com and Webvan.com, out of the wreckage emerged a few notable survivors, including Amazon and eBay. And although the e-commerce model did not translate for most of the dot-com start-ups, it did translate for traditional companies, which were able to unlock hidden value by capitalizing on advances in technology to plug the e-commerce model into their existing corporate infrastructure. And as you know, it is rare to come across a Fortune 500 company today that does not sell its goods or services online.

I'm not worried about Airbnb and Uber, as I think they will emerge as the leaders of the sharing/on-demand economy. But just like with e-commerce, while the marketplace model may not translate for most of the start-ups, it will translate for established companies, for they will be able to unlock hidden value by plugging the marketplace model into their existing corporate infrastructure. Unlike start-ups, established companies have a corporate ecosystem that provides them with the potential to collaborate with their stakeholders to access inventory on the supply side, the necessary customer capital on the demand side, and the structural capital (i.e., tangible and intangible assets) that allows them to better originate, extract, and capture value. And I predict that a decade from now it will be rare to come across a Fortune 500 company today that has not incorporated this form of collaborative commerce into its business model.

The perfect storm we're seeing is creating an opportunity for established companies looking to learn and test out this radical new form of collaborative commerce. By collaborating with sharing/on-demand economy companies through partnerships, strategic investments, or acquisitions, companies will gain an experienced guide to help them navigate the challenges and opportunities of the new white space of the sharing/on-demand economy. And start-ups will gain access to the capital they need in order to survive—not just financial capital but, more importantly, supplier capital, structural capital, and customer capital.

The roots of the marketplace model are much deeper than those of the e-commerce model, though. For while the dot-com movement capitalized on the structural shift in just one force (i.e., technology—the adoption of the World Wide Web and the ability to transact online), the sharing/on-demand economy movement capitalized on structural shifts in the economy, technology, and society. For example, Airbnb built a sophisticated, leading-edge technology platform that capitalizes on people's economic need to earn an alternative form of income and their societal desire to create a more meaningful life. In fact, its very conception, initially called Airbedandbreakfast.com, was born of economic necessity.

Roommates Brian Chesky and Joe Gebbia were struggling to pay their rent. When they heard that San Francisco's hotels were fully booked for an upcoming design conference, they came up with the idea of renting out airbeds on their living-room floor and serving breakfast to their guests. Being digital natives, Chesky and Gebbia were tech savvy and therefore able to put up a website promoting their new business the very next day. After finding their idea to be successful, they reached out to Gebbia's former roommate, Nathan Blecharczyk, a computer science graduate and programmer, to develop the website. Since then, they have been continuously improving the platform by incorporating advances in technology (e.g., smartphones, mobile apps, cashless payment systems, user-generated reviews, social networks). And most importantly, they have successfully capitalized on the structural shift toward social consciousness by creating a global movement for their community-based social mission. As Chesky puts it, "When you share your space with somebody it is a personal, meaningful experience."

By accessing the new social value drivers of advocacy, connection, and collaboration, Airbnb was able to unlock the hidden value of personal capital. In doing so, the company met people's needs and desires, pushing out the boundaries of the consumer value proposition. My own experience with using Airbnb during a business trip to Southern California is a good example of this. Instead of staying at the high-status, five-star (yet cookie-cutter and, to be honest, soulless) business hotel closest to the office, I decided to make the trip an adventure and stay at a mid-century modern house in Laguna Beach. In functional terms, by using Airbnb, not only was I able to access a greater variety of choice, but it was half the cost and more convenient: instead of being stuck in the middle of nowhere at the end of the workday, I was able to go for a walk along the beach. In emotional terms, although it felt a bit strange the first night to be sleeping in the host's bedroom while he slept in the guest room down the hall, it ended up being a great experience, as he was at home the following night and we sat down in his living room, where we drank wine and chatted.

And in psychological terms, by the end of the stay, I really did feel like I was living Airbnb's community-focused mission to "imagine a world where you can belong anywhere," and I was eager for my next Airbnb adventure.[2]

Introducing Ubernomics

Forward-thinking companies are starting to realize they need to enhance their consumer value proposition to meet the higher order and increasingly demanding needs and desires of their customers. For example, Ford's new physical and digital platform, the FordPass, was inspired by the company's desire to emulate the successful, customer-centric strategy of Apple, Nike, Nespresso, Ikea, and Disney, among others. And during its 3Q15 analyst conference call, the president and CEO of Hyatt Hotels Corporation, Mark Hoplamazian, foreshadowed the company's entry into the sharing economy when he made these insightful remarks: "From our perspective... we've, for some time, looked at this whole sharing economy dynamic as a broad consumer issue and the consumer behavioral change, and we've always been drawn towards it, not sort of away from it."[3]

The bottom line is that the convergence of structural shifts in society, technology, and the economy has given rise to a new form of collaborative commerce that allows companies to defy traditional economic principles of scarcity in terms of both supply and demand:

- **Long tail of supply.** As Chris Anderson wrote in 2006 in his book *The Long Tail,* "The theory of the Long Tail can be boiled down to this: Our culture and economy are increasingly shifting away from a focus on a relatively small number of hits (mainstream products and markets) at the head of the demand curve, and moving toward a huge number of niches in the tail."[4] Building upon this, I believe the democratization (or unbundling) of physical and human capital enables companies to access a long tail of latent or underutilized assets, goods, and services. This creates abundance of supply, as inventory growth is not limited by traditional time or capital constraints.

- **Blue ocean of demand.** As W. Chan Kim and Renee Mauborgne wrote in 2005 in their book *Blue Ocean Strategy*, "Blue oceans… are defined by untapped marketspace, demand creation, and the opportunity for highly profitable growth."[5] Building upon this, by meeting functional, emotional, and psychological needs and desires we did not even know we had, companies are able to attract new tiers of noncustomers and access new blue ocean market demand. This creates abundance of demand, as companies are able to expand their total addressable market beyond traditional categories.

As Seth Godin astutely observes in *The Icarus Deception*, we are undergoing a profound structural shift in society and the economy, from one driven by scarcity to one driven by abundance:[6]

> *Scarcity is the dominant driver of the industrial age. Scarce resources, scarce machines, scarce labor and scarce shelf space... The connection economy thrives on abundance. Connection creates more connection. Trust creates more trust. Ideas create more ideas.*

But what does this mean for economics, which is defined as the science of choice under scarcity, and for the field of economics, which is to understand how to fulfill unlimited wants with limited or scarce resources?

I coined the term *copianomics* in September 2014 to describe the science of choice under abundance, as the Latin word for abundance is *copia* (in Greek mythology, *cornucopia* means "horn of plenty"). But this term failed to resonate with my community, so in early 2016 I changed my term of choice to *ubernomics*. And later I discovered that *uber* means "above" in German, which is what this book is all about: the hard-core capitalism of ubernomics for companies looking to rise above the competition. In case it doesn't go without saying, I did not invent the term *ubernomics*—a term that will garner you over 4,900 results when you Google it. There is actually another book titled *Ubernomics* scheduled for publication (and that I'm looking forward to reading), which, according to its author, Jeremy Waite, former head of digital strategy at Salesforce Marketing Cloud and current strategic marketing evangelist for IBM, is a behind-the-scenes look at Uber and at the economics behind the world's most disruptive brand.

It is exciting to imagine the endless possibilities for ubernomics as companies discover that the holy grail of value creation already lies within them. Just as the democratization of content in the past decade led to structural disruption in media-related sectors (just ask the former executives of Blockbuster, Yellow Pages, and Kodak), the emerging democratization of physical and human capital will lead to a radical transformation of the corporate landscape. What I love about ubernomics is that it exposes what I call "the fallacy of the shareholder-centric business model," negating Milton Friedman's traditional scarcity-based doctrine that "there is one and only one social responsibility of business—to use its resources and engage in activities designed to increase its profits so long as it stays within the rules of the game."[7] For ubernomics is about empowering, not exploiting, stakeholders, and changing from a mindset of scarcity to a mindset of abundance. Ubernomics enables companies to originate, extract, and capture value through their existing ecosystem by designing innovative ways to tap into the underutilized and latent physical and human capital of their firm, employees, customers, suppliers, and partners.

Ubernomics is no longer just the domain of start-ups. Industrial Age companies like General Motors, Ford, and Hyatt are starting to make strategic moves into this space, joining the growing ranks of forward-thinking companies like Amazon, LinkedIn, and Expedia. Although these companies have vastly different business models, they all have one thing in common: they recognize the exciting value creation potential of the new white space of collaboration.

2

A Wake-Up Call for CEOs

Shift from Scarcity to Abundance Started with Content

> *We believe investors may be naively assuming directory companies will be able to simply shift their advertising clients from print to online and maintain their current high level of free cash flow. However, we believe the 2006 business book* The Starfish and the Spider *provides a note of caution and a dose of reality. According to the authors, "starfish" organizations, such as Napster, Craigslist, and Wikipedia, which rely on the power of peer relationships, are causing an increasing threat to traditional "spider" organizations, which have weaved their webs over long periods of time, slowly amassing resources and becoming more centralized... We believe one of the authors' key principles of decentralization—as industries become decentralized, overall profits decrease—provides a note of caution for directory companies.*

—"Yellow Pages No Longer a Conservative Investment: Downgrading to SELL," Barbara Gray, CFA, Analyst, Consumer/Diversified, Blackmont Capital, January 10, 2008

I clearly remember going through Rod Beckstrom and Ori Brafman's *The Starfish and the Spider* in early 2008 with my highlighter, searching for insights into the fate of Yellow Pages. Back then, most investors were still steadfast in the long-held belief that directory companies were a source of predictable and sustainable cash flow, resulting in a superior business profile. But I was not your typical follow-the-herd, sell-side equity analyst. Back in September 2006, I had incurred the wrath of a few corporate clients that I classified as "straw houses at risk of collapsing" in my controversial research report *If the Economic Wolf Comes Knocking, Is Your Trust Built of Bricks or Straw?*

Having used Craigslist to find a place to live, and from my ongoing conversations with my then boyfriend (now husband), Greg, about how technology was disrupting traditional business models, I knew I could no longer be in denial. It was Beckstrom and Brafman's book that really crystalized it for me and gave me the conviction to go against the consensus and downgrade Yellow Pages to sell. Four years later, Yellow Pages' stock traded down to the pennies, and the company filed to restructure.

As we witnessed during the past decade, the democratization of content led to structural disruption of a wide range of media-related sectors. As Jeff Jarvis shares in *Public Parts,* which I read back in October 2011, "'The large profit margins newspapers enjoyed in the past were built on an artificial scarcity: Limited choice for advertisers as well as readers,' Google said in a paper submitted to the U.S. Federal Trade Commission in 2010. 'With the Internet, that scarcity has been taken away and replaced by abundance.'"[8] And now we are seeing a similar shift from scarcity to abundance with the emerging democratization of human and physical capital, which will lead to structural disruption over a much wider range of sectors. And as foreshadowed by the plummeting price of taxi medallions, reported in the *New York Times* in its November 2014 article "Under Pressure from Uber, Taxi Medallion Prices Are Plummeting," the dominoes have started to fall. Here is one particularly insightful excerpt from that article:[9]

> *In major cities throughout the United States, taxi medallion prices are tumbling as taxis face competition from car-service apps like Uber and Lyft... The average price of an individual New York City taxi medallion fell to $872,000 in October, down 17 percent from a peak reached in the spring of 2013, according to an analysis of sales data. Previous figures published by the city's Taxi and Limousine Commission—showing flat prices—appear to have been incorrect, and the commission removed them from its website after an inquiry from* The New York Times... *In other big cities, medallion prices are also falling, often in conjunction with a sharp decline in sales volume. In Chicago, prices are down 17 percent. In Boston, they're down at least 20 percent, though it's hard to establish an exact market price because there have been only five trades since July. In Philadelphia, the taxi authority recently failed to sell any medallions at its asking price of $475,000; it will try again, at $350,000.*

Since then, the price of New York City taxi medallions has continued to tumble. In August 2016, they were listed for sale on NYCityCab.com for as low as $300,000 to $500,000. I came across evidence that this once highly valued asset is now viewed as a liability—a February 2016 sales listing offering a NYC taxi medallion for $0, with this explanation: "Moved out of NYC. Need someone to take over current payments on medallion and the medallion is yours." Interestingly, in May of that year, Medallion Financial Corp., which used to describe itself on its website as a "specialty finance company with a leading position in the origination and servicing of loans financing the purchase of taxicab medallions and related assets,"ominously changed its stock symbol from $TAXI to $MFIN. Investors would have been wise to heed this red flag, as less than three months later, in August, the company cut its dividend by 80%.

Ubernomics: The Radical New Form of Commerce

Uber is having a highly disruptive impact on the taxi industry, and what's incredible is that, in the seven years following its inception, the company expanded to having a presence in over 400 cities globally. In my search to understand how Uber is able to achieve such an accelerated rate of disruption, I came across a 2004 *Harvard Business Review* article by Anita McGahan titled "How Industries Change."[10]In it, she observes:

> *Radical transformation occurs when both core activities and core assets are threatened with obsolescence. The relevance of an industry's established capabilities and resources is diminished by some outside alternative; relationships with buyers and suppliers come under attack; and companies are eventually thrown into crisis.*

McGahan goes on to say that radical industry evolution is "relatively unusual" and that an "industry generally evolves along just one trajectory at a time." Uber is totally radical, emerging from left field like it did as a disruptive threat both to the taxi industry's core activities (it poses a more attractive alternative to drivers and passengers) and to its core assets (the premium-priced taxi medallions, which create artificial scarcity). I don't think it will take a whole decade for the change to play out. Uber is achieving an unprecedented form of accelerated disruption, as the switching costs for both suppliers and buyers are low (the only obstacle being regulatory constraints). In her article, McGahan points out that radical change usually occurs after the mass introduction of some new technology. To understand the technological forces that gave rise to the emergence of Uber, we need to go back to the early days of the Internet.

We started off surfing the Internet in the early 1990s with the advent of the World Wide Web. In 1995 we rode the rising dot-com wave, dipping our toes into the ocean of corporate e-commerce, until the wave came crashing down in March 2000. Out of the corporate dot-com wreckage emerged the next generation of companies that brought us user-generated content (Blogger in 1999, TripAdvisor in 2000, Yelp in 2004, YouTube in 2005) and social networking (LinkedIn in 2003, Facebook in 2004, and Twitter in 2006). These companies created a new undersea social world of transparency, authenticity, and engagement, promoting individual empowerment, community, and collaboration. More importantly, these web 2.0 companies created online networks of people. And over the span of a decade, their membership bases grew, new social exchanges emerged, and the density of connections within and between members of the different social networks increased. This led to the foundation of a massive global network of highly connected individuals, laying the groundwork for the emergence of one of the most diverse and self-sustaining ecosystems on earth.

In her *Harvard Business Review* article, McGahan speaks of how radical change can occur as consumer tastes change, which, in addition to the mass introduction of new technology, explains how, in the aftermath of the Great Recession that occurred in the late 2000, the sharing economy movement started to gain momentum because of structural shifts in these three forces:

- **Technological.** The ability for companies to create platforms for people to easily transact among themselves was enabled by the mass adoption of the smartphone (which emerged with the June 2007 introduction of the Apple iPhone) and the creation of mobile apps (starting a year later with the launch of Apple's iPhone App Store, in July 2008). Through leveraging all the recent advances in technology (i.e., online marketplaces, cashless payment systems, user-generated reviews, social networks, smartphones, mobile apps), sharing economy companies are creating radical changes by offering a superior functional value proposition to sellers and buyers of their products or services.
- **Economic.** As individuals recovered from their excessive debt-fueled individualistic and materialistic consumerism-binge hangover from the Great Recession, they started to seek more meaningful lives and alternative means to earn an income. Companies can meet this demand when they create a unique, authentic, and personal experience for sellers and buyers, enabling a strong emotional connection to their product or service. From an economic perspective, sharing economy companies enables sellers to earn passive income (by renting out their assets or selling their used goods) and active income (through monetizing their own human capital) while enabling buyers to achieve access without ownership.
- **Societal.** The wave of socially conscious and empowered

millennials will be an influential force on corporate America and Wall Street over the next few decades as they seek to align their values with those who they buy from, work with, do business with, and invest with. As evidence of this shift to social consciousness, in September 2015, Forbes published its first ever "Change the World" list, which interestingly ranks companies not by the dollars they make but by the good they are doing. As the new generation of sharing economy companies is especially social mission–driven, these companies are well positioned to meet this demand. And, more importantly, these companies are starting to build thriving stakeholder ecosystems as their sellers, buyers, partners, and employees form a psychological attachment to what these companies stand for. As Robert J. Shiller writes in *Finance and the Good Society*, "Capitalism has to be expanded and democratized and humanized."[11]

Expectations Are Rising in the Age of Uber

I experienced one of the most frightening times of my life in September 2014, when I found myself grabbing my iPhone and dialing 911 while I watched in panic as my then 15-month-old baby, Adam, lay convulsing in my husband's arms. As I breathlessly repeated our address for the third time, my heart racing, I thought, *Why can't I just push a button to send for an ambulance?* And as Greg and I stood with tears in our eyes, waiting helplessly for the ambulance to arrive for Adam, who lay limp and pale, I thought, *Why can't I just look at my iPhone and track the ambulance to see how far away it is?* It was an extension of the way my thinking had transformed since six months prior, when Greg and I had traveled to New York City and experienced the magic of Uber for the first time. After that ride, I resolved never to set foot in a taxi again, and I found myself growing more and more impatient with how archaically the rest of the world seemed to operate.

I don't understand why all service providers can't provide such a seamless, enjoyable, and convenient customer experience as Uber does. As Greg will attest to, I got into a heated debate with our moving company in May 2015 when I called to find out when we could expect the crew and was informed they would arrive "between 7 a.m. and 10 a.m., depending on traffic." "Seriously?" I responded. "Can you not just use Google Maps or Waze to give me a closer estimate of the arrival time?" And as we waited impatiently for the moving truck to pull up in front of our house, I kept thinking, *Why can't I track the moving truck on an app? Why do I have to deal with the annoying impersonal dispatch person in Florida, instead of just communicating directly with the guys in the truck?* This frustration would be repeated weeks later when we were settled in our new house. I became increasingly irritated with being put on hold while on the phone to the various utility companies—and then having to wait around all morning for their workers to show up.

As an equity analyst, I can't help but wonder how legacy service companies will be able to compete in the new on-demand era. Will they end up like the Maytag Man, sitting idly by the phone, waiting for it to ring? Their challenge is that people no longer want to pick up the phone only to be put on hold by a company's seemingly faceless service department or dispatch call center, and to then learn that no one is available to come out until the following week. In the age of Uber, people want to just push a button and have the Maytag Man arrive at their doorstep the next hour, a smile on his face. In the age of Uber, customers will become only more and more demanding. Therefore, the question is not *if* legacy companies will join the on-demand economy but *when*.

Uberized Consumer—A Company's Worst Nightmare

In the fall of 2015, I met up for coffee with Andrew Reid, the founder of Vision Critical, a customer intelligence software firm he launched in 2000 that built upon the foundation of his father's legacy market research company, Angus Reid & Associates. I was excited to meet with Andrew, as his firm had recently published *The New Rules of Collaborative Economy,* a market research study I had recently learned about from the coauthor and founder of Crowd Companies, Jeremiah Owyang, when he presented the key findings of the study at the Collaborative Economy Conference. At the conference, I also met Carey Lefkowitz, a managing director at Bovitz, a Los Angeles/ New York–based marketing research and strategy consulting firm, who presented on the insights from his firm's study *What's Next for the Collaborative Economy: A Consumer-Driven Perspective on Opportunities for the Collaborative Economy.* Interestingly, both market research studies reveal that price and convenience are the key drivers for people to try to participate in the sharing/on-demand economy. When I sat down and talked face to face with Andrew about price, convenience, and brand, I had a flashback to the research I did back in 2010 on advocacy as a new social value driver. All of a sudden, the dots started to connect.

Airbnb and Uber are so disruptive to their respective industries because they meet functional, emotional, and psychological needs and desires that we the consumers did not even know we had. And in doing so, these companies are subconsciously raising our consumer mindset and expectations. For example, on a recent business trip in Southern California, I didn't rent a car or call a taxi to commute between the office and the Airbnb place I was staying—I used Uber. In functional terms, not only was it much cheaper and incredibly more convenient, but I got to experience a wide range of vehicles (e.g., Mercedes, BMW, and Toyota Prius) and drivers (interestingly, most of the people were professionals and used Uber to earn money in order to commute to work, to supplement their income, or to generate real estate or marketing leads). In emotional terms, I looked forward to the commute, as it was fascinating to talk with the drivers and learn their stories. And in psychological terms, I became a strong advocate of Uber's accessibility-focused mission to "bring transportation as reliable as running water to everyone, everywhere." However, as I live in a city that, at the time of writing, disallows ridesharing, I unfortunately only get to experience Uber when I travel. (Note to the City of Vancouver: Is it not your vision to be the greenest city in the world by 2020?)

The real disruptive threat caused by companies like Airbnb and Uber will come from the reactions of people like you and me. I know I can't be the only one who has been looking at the consumer world differently ever since experiencing the wonders of Airbnb and Uber. The appeal isn't just in having the ability to push a button and have a particular good or service delivered immediately; it also has a lot to do with basic humanity—the way we are treated by the person providing the service. Don't we all want the employees we interact with to be as authentic and engaging as the average Uber driver or Airbnb host?

What I love most about social media is that it has empowered us as consumers by giving us a voice—and a very loud one at that. So instead of silently lamenting *Why can't that company be more like Uber?,* we can (and will) join together and use our voices. Smart companies already recognize this and are working to adapt in order to meet the changing needs of their consumers, lest they be left in the dust. While some companies may view the uberized consumer as their worst nightmare, the truly progressive and innovative companies see the opportunity to embrace us.

No Longer Just the Domain of Start-Ups

Now that its potential to create value is better recognized, ubernomics is no longer the domain of start-ups only. In the fall of 2015, technology leaders such as Amazon, LinkedIn, and Expedia announced strategic moves into ubernomics. At the end of September 2015, Amazon launched Amazon Flex, its on-demand delivery service. This goods delivery platform will enable Amazon to cost-effectively build a long tail of delivery drivers for its one-hour Amazon Prime Now service, while creating opportunities for people to easily monetize their underutilized human capital by using their own vehicles and Amazon's routing app. And less than two weeks later, Amazon entered the trades services vertical with the launch of Handmade, a rival to Etsy, featuring over 800,000 factory-free and handmade products from 5,000 sellers. Although Etsy's social mission to "reimagine commerce in ways that build a more fulfilling and lasting world" is admirable, the company must now be questioning its October 2013 strategic decision to broaden its definition of handmade. Although opening its platform to sellers using employees and manufacturing partners enabled it to scale its supply base, it created an opportunity for Amazon to leverage its structural asset base and attack the artisan marketplace.

On October 19, 2015, LinkedIn quietly launched LinkedIn ProFinder, its professional services freelancer marketplace. LinkedIn started piloting it in San Francisco with three categories (accounting, graphic design, and writing and editing); it now includes US-based professionals with expertise across a range of 10 services (including legal, real estate, and financial services). To quickly scale supply and demand, LinkedIn is offering its platform for free, but I'm guessing that within a few years it will be in the position to start taking a cut.

On November 4, 2015, online travel company Expedia entered the sharing economy with its US$3.9 billion acquisition of HomeAway. HomeAway was a pioneer in the personal asset–sharing vertical, launching pre-recession (in 1Q15) and raising a total of US$505 million in venture capital financing before going public at a US$2.15 billion valuation in June 2011. Although the stock rose by over 75% from its US$27 initial public offering (IPO) price to peak at US$47 in late February 2014, it fell back to its IPO price given concern of increasing competition from Airbnb. The acquisition makes strategic sense for Expedia: in October 2013, the two companies entered into a partnership to surface HomeAway's rental properties on Expedia.com. While this acquisition fulfills HomeAway's mission to "make every vacation rental available to every traveler in the world," my money is on Airbnb given its strategic community-focused aspiration to "imagine a world where you can belong anywhere."

As I discuss in more depth in Chapter 10, Industrial Age stalwarts like GM, Ford, and Hyatt have also started to embrace ubernomics. In early January 2016, GM announced a US$500 million strategic investment in Lyft and acquired the car-sharing platform Sidecar a few weeks later. On January 11, at the North American International Auto Show, Ford proclaimed it no longer saw its future as just an auto company but as an auto and mobility company, and announced the April launch of FordPass, its new digital and physical marketplace platform. Likewise, Hyatt recently launched its Unbound Collection, a kind of Airbnb of independent boutique hotels. In each of these cases, "old economy" companies have identified established strengths to which they are now applying "new economy" opportunities analysis. And I expect we will see more and more companies seize the opportunity to start shifting their business models from the traditional era of scarcity to the new era of economic abundance.

3

Defying Economic Principles of Scarcity

Airbnb Has Masterminded the Ultimate Game of Monopoly

Remember playing Monopoly when you were younger? It was one of my favorite board games growing up. Even though I didn't know anything about economics or investing back then, there was something addictive about the game. I especially relished the special moment when I got to trade in the little, green plastic houses for a big, shiny, red hotel. This aspect of the game serves as an excellent model for comparing the economic characteristics of Airbnb with its competition, the traditional hotel industry.

As I note in my in-depth research report published in September 2014, *The Abundance Economy: Where the Long Tail Meets the Blue Ocean*, it took Airbnb only six years to amass an inventory of over 800,000 accommodation listings in 190 countries. In contrast, it took the hotel conglomerate Hilton Worldwide nearly a century to grow to 11 brands operating 4,200 hotels in 93 countries and offering a total of 690,000 hotel rooms. From the summer of 2014 to the summer of 2015, Airbnb attracted an additional 700,000 accommodation listings—more than 16 times the only 42,000 additional rooms Hilton Worldwide gained during the same period. The difference between the two is stark.

Crucial to its success, Airbnb is converting its long tail of inventory into actual revenue. According to a June 2015 *Wall Street Journal* article, "The Secret Math of Airbnb's $24 Billion Valuation," Airbnb's revenues approached US$1 billion in 2015 and were forecasted to climb to US$10 billion by 2020! Based on information shared by the company, the average Airbnb listing is available 66 days of the year. In simple terms, this implies that the average listing on Airbnb is available two months out of every year, or that it takes six Airbnb listings to equate to one hotel room. If we multiply this by the average of 164 hotel rooms per chain hotel, we can calculate that it takes approximately 1,000 Airbnb listings to create the equivalent of one hotel. Think back to Monopoly and the thrill of trading in five little, green houses for a big, red hotel. This is essentially what Airbnb is doing—in this case, trading in 1,000 little, green houses (i.e., 1,000 accommodation listings) for one big, red hotel.

Continuing with the Monopoly-Airbnb comparison, there are further interesting similarities worth looking at. The board game was debuted by Parker Brothers during the heart of the Great Depression, in February 1935; Airbnb launched its real-life online version in the middle of the Great Recession, in August 2008. By 1937, over 6 million copies of Monopoly had been sold—Americans loved being able to play a game that afforded people a chance to win and feel rich during a time when everybody was feeling anything but. Three years after its launch, Airbnb celebrated its one-millionth booking as people welcomed the opportunity to earn extra income by opening up their homes to visitors. Travelers loved the superior value for money, the rich variety of unique places to stay, and the personal experience of staying in someone's home.

The perks of the game are starting to catch on: in the span of seven years, Airbnb has attracted a long tail of 2 million little, green houses, the equivalent of 2,000 hotels. The magic of Airbnb's asset-light business model is its lack of required development capital to build these "hotels." Based on the actual hotel industry model, in which the average cost to build an economy hotel is US$14 million, Airbnb would be looking at a cost of approximately US$28 billion. Given that as of August 2016 Airbnb has raised a total of US$2.4 billion, we know it cannot have spent close to US$28 billion building its game platform.

To reach its ambitious target of US$10 billion in revenue from the US$900 million it was estimated to have earned in 2015 (the exact number has yet to be reported at time of writing), Airbnb will need to grow its accommodation listings by 11 times (assuming there is no change in its commission rate, average room rate, or average days of availability). This implies a long tail growth in accommodation listings from 1.2 million (at the midpoint of 2015) to 13.2 million. Over the next five years, Airbnb is looking to add over 12 million little, green houses to its online version of Monopoly, the equivalent of adding 12,000 big, red hotels, which I calculate would cost a traditional player like Hilton US$170 billion. From a customer standpoint, assuming there is no change in the average nights booked, this would mean that Airbnb expects to grow its player base by 11 times from an estimated 16 million in 2015 to over 175 million players by 2020.

Can you imagine Hilton's next analyst conference call if management announced it was planning to spend US$170 billion over the next five years to add 12,000 hotels to its current base of 4,300? Obviously, this would be an impossible feat for Hilton, since it is playing a more traditional game of Monopoly in an environment of scarcity (i.e., scarce land, scarce capital, and scarce resources).

Just like you can't put up a red plastic hotel on Boardwalk with the first roll of the dice, Hilton cannot open up a new hotel on Broadway overnight. First it needs to find the right property for development (i.e., wait to land on the property itself), buy the land (i.e., pay the bank to get the property deed), and then spend one to two years developing the hotel (i.e., wait each turn to buy a house).

This is where Airbnb has a distinct advantage: whereas Hilton operates in the traditional era of scarcity, Airbnb operates in the new era of abundance. Airbnb is not just playing by different rules, it has become the mastermind in the ultimate game of Monopoly, and people all over the world are starting to come together to play.

Uber Expands Its Total Addressable Market beyond Traditional Categories

When Uber CEO Travis Kalanick spoke at the Digital Life Design conference in Munich, Germany, in January 2015, he claimed that the taxi market in San Francisco was generating about US$140 million in driver revenue per year. He followed this by sharing that Uber's driver revenues in San Francisco, meanwhile, were running at the comparatively astounding amount of US$500 million per year.

The reality is, companies like Uber are not only disrupting the incumbents but also gaining access to new tiers of noncustomers and opening up a new market of nonconsumption, leading to an abundance of demand. It's amazing to think that, after operating in San Francisco for only four years, the company was generating revenue in excess of three times that of the traditional taxi market.

Uber's accessibility-focused mission to "bring transportation as reliable as running water to everyone, everywhere" expands its total addressable market (TAM) beyond the traditional taxi market to noncustomers accustomed to taking public transport and renting or owning vehicles. By this same logic, Airbnb's highly inspirational community-focused social mission could expand its TAM beyond the traditional lodging category to include noncustomers who used to stay with friends or relatives when traveling, signed annual leases to rent a house, or owned vacation homes or timeshares.

According to the American Hotel & Lodging Association, the traditional US lodging industry (e.g., hotels, motels, timeshares, B and Bs, hostels) generated US$176 billion of revenue in 2015.[12] And global hotel industry revenues are forecasted at US$550 billion for 2016, over three times this amount. If we divide US$91 billion (the gross bookings that Airbnb would need to generate in 2020 to meet its US$10 billion revenue projection) by US$550 billion, this implies it would gain market share of just under 20%.

One way Airbnb is already expanding the traditional travel market is by providing the opportunity for people to experience new places as a local rather than as a tourist. As Peter Thiel, cofounder of Founders Fund, the lead investor in Airbnb's US$200 million Series C round in October 2013, wisely asserts in the book he coauthored with Blake Masters, *Zero to One*, "Creative monopolists give customers more choices by adding entirely new categories of abundance to the world."[13]

It Takes a Movement, Not Just a Marketplace

The sharing/on-demand economy universe is rich with founders who epitomize the words Jonah Sachs uses to describe "The Rebel" in his book *Winning the Story Wars*: "Creative destruction of the status quo… idealistic vision of a better way… fearless, uncompromising, and creative."[14] In this sense, you can consider founders like Brian Chesky and Travis Kalanick as being rebels with a cause who have created not just marketplaces but entire movements. And as Scott Goodson observes in *Uprising*, movements are what attract people who are "hungry for meaning, authenticity, sense of belonging, and purpose… [and] beginning to engage with and shape culture around them as opposed to being passive consumers of culture created for them by others."[15] It is the movement, and not just the marketplace, that enables Airbnb and Uber to defy traditional economic principles of scarcity and operate in a new era of economic abundance.

It is fascinating to watch Airbnb and Uber as they continuously innovate. For example, in October 2015, Airbnb announced the test launch of Journeys, a fully managed travel service. This strategic move has the potential to expand Airbnb's TAM beyond the traditional category of accommodation by allowing people to experience curated excursions with locals through democratizing the latent skills and passion of their existing host community. It hints at Brian Chesky's ambition to evolve the company into a full-blown hospitality brand that provides a reimagined, end-to-end seamless travel experience and further advances his highly inspirational community-focused social mission.

A month after Airbnb's test launch, Uber announced a new offering it would be piloting in the San Francisco Bay Area. Called Driver Destinations, the service makes commuting more affordable and convenient by enabling its drivers to earn extra income by picking up passengers on their drive to and from work. This strategic move, which is completely aligned with Uber's accessibility-focused mission, further pushes out the boundaries of the consumer value frontier and further expands its TAM. As Kevin Manley observes in his book *Trade-Off* (which, ironically, I read back in early 2009, right around the time Airbnb and Uber were being launched), "Technology constantly improves both fidelity and convenience."

Judging from the fact that the top 75 North American sharing/on-demand companies raised US$9.6 billion in 2015, double the US$4.8 billion raised in 2014, venture capital investors have also been excited about ubernomics. Although most of my research has been focused on Uber, I have also been rooting for the underdog, Lyft, which seems to be its nicer and humbler cousin. I was inspired by how the company, which evolved in May 2012 out of Zimride, a ridesharing service company launched in 2007, has stayed true to its sustainability-focused social mission to "take cars off the road, not replacing or augmenting existing systems." I remember the earnestness and passion of Logan Green, cofounder of Lyft, as he delivered his keynote speech "Fixing Transportation with Humanity and Technology" at SXSW in March 2015, sharing his vision for the world of making car ownership unnecessary.

So I was pleasantly surprised when I learned just before the end of 2015 that Lyft raised US$1 billion, more than doubling its valuation from US$2.5 billion to US$5.5 billion—but I was shocked to discover the lead investor was General Motors. I struggled to understand why a company whose vision is to eliminate vehicle ownership would enter into a strategic alliance with the world's largest vehicle maker. As I watched the CNBC interview with Daniel Ammann, president of General Motors, on the morning of the announcement, I was puzzled to see Logan Green enthusiastically endorsing their new alliance. It is a brilliant move on GM's part since, in addition to providing it with a new distribution channel for its vehicles, it enhances its brand equity and gives it a seat at the table of a company trying to disrupt its vehicle sales business. And it makes sense that Lyft sees autonomous-driving vehicles as the future, but would not Google or Tesla be a better cultural fit for it? It also seems ironic that part of the deal is for GM to supply vehicles under a hybrid rental/leasing contract to people who want to drive for Lyft but whose existing cars do not qualify. And from a long-term strategic perspective, would Lyft not be better off following the lead of Uber, which in July 2015 launched XChange Leasing, a pilot program whereby Uber directly leases used and new vehicles to its drivers from a wide range of manufacturers (e.g., GM, Toyota, Ford, Hyundai, Nissan, VW, and Chrysler).

When I wrote about this back in early January 2016, I speculated that Lyft was hoping to avoid the fate of Sidecar, which had just shut down its ridesharing and delivery operations after raising US$40 million since its launch in January 2012. I expressed hope that this new injection of capital would provide Lyft with the resources to continue to scale its platform and compete against Uber. And, ironically, a few weeks later, GM announced it was also acquiring Sidecar. I remain concerned that this could be a red flag that Lyft is losing its "rebel with a cause" attitude, as this is what inspires people (both drivers and passengers) to join the Lyft movement. But what I have now realized is that by bringing in GM as a strategic investor, Lyft is gaining access not only to financial capital but to supplier capital, which could be critical as competition with Uber intensifies.

II

The Three New Social Value Drivers

4

The Value of Advocacy

Social Media Democratizes Influence

As I reviewed the six years' worth of research reports and articles I'd written, I decided to also dig through my old journals and notebooks. I came across this journal entry:

> *February 2, 2010—So, I am at a crossroads trying to figure out what to do. I feel the need or desire to create something. Add real value to the world rather than just work for a firm. I need to find something I can be totally passionate about that would fulfill my talents and strengths so I can get back in the flow. But the question is, What? When people ask me what I am doing and I tell them I resigned from my position as an equity analyst at an investment firm, I think they are confused as to why. But in my mind, I need to create this gap so I can come up with the next big thing to carry my career/life forward in a meaningful way.*

My decision to take a break in order to find meaning in my life—including my career—would lead me, one month after writing that journal entry, to embark on a two-and-a-half-month road trip across the United States with Greg, who was then my new fiancé. During that trip we attended the SXSW conference in Austin, which is what opened my eyes to the exposing disruptive force of social media and, using Joseph Campbell's terms, summoned my call to adventure. It was the starting point of my "hero's journey" into the emerging land of abundance economics that was being shaped in real time by the convergence of rapid structural shifts in societal, technological, and economic forces.

But I was not yet ready to heed the call to adventure, as the status and comfort of working for an investment firm beckoned strongly to me upon my return home. But then, a few weeks later, I discovered I was pregnant with my first child. I was 39 years old, and I decided not to go back to the corporate world—mainly because I was worried the stress I'd experience might harm the baby. But once again I refused the call and instead signed up for a three-month real estate course—my nesting instinct had set in, and I figured it might help Greg and me to buy a house. But once I'd finished the course, I realized I had lost my sense of identity, as my journal entry from that time describes:

> *August 28, 2010—I am really dealing with an identity crisis. Although my objective since I moved back from NYC was to find a great guy to settle down with and start a family, now that I am there, I am missing my old life... I'm reading this book called* Wander Woman *and it is making things clear for me. According to the book, my six archetypes are Wanderer, Adventurer, Thinker, Teacher, Superstar, Visionary. And whereas I used to thrive on these with my career and lifestyle, I am now constrained and not fulfilling any of them. So*

> *that explains why I am feeling so bored and unfulfilled. I am beginning to question why I want to buy a house on the North Shore. If I am planning to resume my career, would it not make more sense to just get a condo?... Why do we need to have a backyard and be by schools when we have five years until the kid needs to go to school?*

My reading the wisdom that the author of *Wander Woman*, Marcia Reynolds, shares turned out to be a crucial turning point in my journey. It gave me the insight I needed and the confidence to start putting my analytical and research skills back to work.

In early January 2011, shortly after the birth of my baby, who we named Brady, I published my first research report as an independent analyst, *Social Media: An Exposing Disruptive Force*. This somewhat controversial report marked the crossing of the threshold for me (again using Joseph Campbell's terminology). I departed from the Ordinary World into the Special World—starting my deep descent into the Underworld of Capitalism. As Gary Klein states in *Seeing What Others Don't*, "New ideas often emerge from a breakdown of previous beliefs. New paradigms emerge from the recognition of anomalies and contradictions."[16] It was my radical thesis that social media was emerging as an exposing disruptive force that laid the foundation for what would evolve years later into my thesis about ubernomics, a new paradigm for economics. Unknown to me at the time, I had just discovered the emerging social value driver of advocacy.

Looking back through that report five years later, I was reminded of the biggest foreshadower of ubernomics. It was something Mark Zuckerberg said on November 16, 2010, at the Web 2.0 Summit in San Francisco: "Facebook will be an 'enabler' for many companies that will disrupt verticals by introducing social functionality over the next five years." At the time I wrote my report, I had interpreted Zuckerberg's statement only insofar as it reinforced my social capital thesis that investors would no longer be able to rely on traditional valuation metrics since they would need to find new ways to start looking at and analyzing consumer-facing companies. If only I'd had the foresight then to think about his words in the context of the nascent sharing economy as it was explored in two books I had just read: Lisa Gansky's *The Mesh: Why the Future of Business Is Sharing* and Rachel Botsman and Roo Rogers' *What's Mine Is Yours: The Rise of Collaborative Consumption.* But it would be another four years until I would connect those dots and realize that social functionality would lay the groundwork for the mainstream emergence of the sharing economy.

My basic investment thesis was that investors would need to start looking beyond the numbers to the depth and authenticity of a company's customer relationships. Back in 2010, we were starting to witness a structural shift in consumer behavior from passive to active as the then recent advent of web 2.0 and social media platforms provided consumers with the ability to conduct their own due diligence on companies. Instead of simply relying on and trusting traditional advertising claims by companies, consumers were actively seeking out and sharing the truth about companies and their products and services.

Although the social media data for companies (in terms of their number of Facebook "likes" and Twitter "followers") was changing daily, I believed investors could gain valuable insight into consumer-facing companies by starting to incorporate social media metrics in their due diligence process. Interestingly, when I attended the Interactive Conference at SXSW in Austin and the 140 Character Conference in New York, both in 2010, the financial community was noticeably absent. And in speaking with my former investment industry colleagues on both the buy side and the sell side, it appeared that most of the Street was not yet looking at companies' social media efforts. I believed that, with Goldman Sachs' then recent decision to invest in Facebook, it would not be long before the two domains of social media and investing came together, but until then, investors had a unique window of opportunity to gain an edge over the competition.

One of the presenters at the 2010 SXSW was Frank Eliason, the former customer service manager at Comcast who is legendary in social media circles for being one of the early corporate embracers of Twitter and for coming up with the idea of using Twitter to interact with Comcast customers. I still remember how he mentioned in 2008 that only a handful of companies were active on Twitter—Starbucks, Southwest, JetBlue, and Dell.

My theory at the time was that, over the next few years, social media would emerge as the core value accelerator, creating three classes of consumer-facing companies:

- **Heart and soul.** The greatest investment opportunity existed in identifying the companies with true "heart and soul." These select companies, which had authentic and deep customer relationships, would be ideally positioned to leverage both the high level of enthusiasm for their brands and their grassroots community marketing efforts through social media, resulting in a positive multiplier network effect.
- **Heart.** Companies with "heart," those whose customers have a deep emotional attachment to their brands, were positioned to leverage this high level of enthusiasm through social media, resulting in a positive network effect.
- **Shell.** Companies with just a "shell," those whose customers had no real emotional attachment to their products or

> services and viewed them as merely utilitarian or convenient in nature, would not be successful in their attempts at social media. Investors would want to red flag and avoid or short these companies, as social media would expose the inauthenticity and shallowness of their relationships with customers, employees, and other stakeholders.

I conducted a case study on six consumer-facing companies I had previously researched, to illustrate how investors could use my social media analytical framework. I won't bore you with the details, but based on my quantitative and qualitative analysis, I classified lululemon and Starbucks as companies with "heart and soul"; Coach, Tiffany & Co., and Tim Hortons as companies with "heart"; and Yellow Media as a "shell" company.

I'd had a long history with lululemon, as I had initiated coverage on the company on September 11, 2007, shortly after its IPO, with a buy recommendation based on my investment thesis that the company had created a unique "blue ocean strategy" and was positioned for superior sustainable earnings growth. Part of my conviction also came from the parallels I saw between lululemon and Starbucks after reading *Pour Your Heart into It*, in which coauthor Howard Schultz, CEO of Starbucks, shares that the company's goal was to "build a great company—one that stood for something—one that valued the authenticity of its product and the power of its passion."[17] And Christine Day, who joined lululemon as president in early 2008, was a 20-year Starbucks alumna and one of Schultz's first hires.

However, back then, many investors dismissed this "girlie yoga" stock as a high-priced fad, believing nobody would pay US$90 for a pair of yoga pants, no matter how good they were. And from a valuation perspective, they balked at its astronomical price-to-earnings multiple of 100 times. But I believed that if you peeled back the company's layers, you would get a much deeper insight into the company. Building upon my heart and soul investment thesis, I visualized lululemon as having the following three layers:

- **Shell.** The basic outer layer of a company is its "shell," which represents the utility value proposition (financial and functional elements) that its product or service offers. lululemon has a solid "shell"; although its apparel is premium-priced, this is more than offset by its long economic life and the high quality of both the company's product and service.
- **Heart.** Some companies go one level deeper and have "heart," which represents the emotional connection people have with the company's brand. lululemon has "heart," since it has created a strong emotional attachment by offering its customers the promise of a more active, happy, and healthy lifestyle, and the brand symbolizes being fit and active—and looking good while doing it.
- **Soul.** The companies with the deepest level of stakeholder relationships are the ones with "soul," which represents the strong psychological attachment that people have with the company's greater purpose. What differentiates lululemon from most companies is its "soul": its mission statement is "creating components for people to live longer, healthier, more fun lives," and its vision is "elevating the world from mediocrity to greatness." The company's culture was born from a group of yogis, dancers, runners, and customers who follow the brand religiously and act as volunteer evangelists.

Coincidentally, just after my report was published, I read Simon Sinek's brilliant *Start with Why*, and I discovered that my shell, heart, and soul concept closely aligned with Sinek's golden circle concept of who, what, and why. As he observes in his book, "Knowing WHY is essential for lasting success and the ability to avoid being lumped in with others. It is the cause represented by the company, brand, product, or person that inspires loyalty."[18]

Despite lululemon being a retailer of highly discretionary, premium-priced athletic apparel operating in a recessionary environment, its stock price was up nearly 100% in the fall of 2011 from when I initiated coverage on it four years earlier. I believed the increase in the firm's valuation was a function of an increase in its growth rate (its earnings per share more than quadrupled from US$0.36 in 2007 to US$1.69 in 2010) and a decrease in its discount rate (its economic moat widened as the company's grassroots community marketing program created a positive network effect). This simultaneous increase in the company's growth rate and decrease in its discount rate expanded its core soul, resulting in a positive multiplier effect with the simultaneous expansion of its heart and shell layers.

On the other end of the customer engagement spectrum was Yellow Pages (or rather, Yellow Media, as it had recently changed its name to at the time), which exemplifies what happens to a company with an empty shell that starts to crack. As mentioned in Chapter 2, in January 2008 I went against the Street and downgraded the stock to sell, based on my investment thesis that Yellow Pages' formerly strong economic moat was narrowing and we would start to see customer attrition as its traditional directory business model was disrupted. However, many on the Street remained in denial about these structural changes, maintaining their conviction in the company's strong brand name and customer relationships. But over the ensuing three years, the company's shell had contracted, with its stock price declining by over 50%. I believed this erosion in the firm's value was a function of a decline in its growth rate (earnings per share growth fell short of analyst expectations) and an increase in its discount rate (its economic moat narrowed as disruptors lowered switching costs for its small-and medium-sized enterprise customers). This simultaneous decrease in the company's expected growth rate and increase in its discount rate resulted in a contraction of its shell.

I warned investors to be wary of companies like Yellow Pages that had decided to hide behind their corporate logos—I saw the company as an empty shell trying to cover up serious customer relationship issues that could be exposed by social media. The reality is that many of these empty-shell companies had been able to get away with offering an inferior value proposition because their customers had limited alternatives due to restricted competition resulting from monopolies, government regulation, or high switching costs. However, as advancements in technology drove the cost of marginal distribution to zero for many industries, these companies saw their traditional business models disrupted as new entrants entered the market and lowered switching costs. But many of these traditional companies were still in denial, believing their customers would stay with them forever and that they would be able to retain their brand glory. However, in truth, these empty shells were at risk of cracking since industry deregulation and disruptors posed a significant threat of substitutes. And my thesis was that social media would start to expose the lack of authenticity and shallowness of these companies' relationships with their customers, employees, and other stakeholders.

Advocacy Creates Abundance of Demand

One of the most influential mentors I encountered early in my journey was John Mackey, founder and co-CEO of Whole Foods Market. In March 2011, I came across a blog Mackey wrote back in 2006 titled "Conscious Capitalism: Creating a New Paradigm for Business," in which he proposes, "Having a deeper, more transcendent purpose is highly energizing for all of the various interdependent stakeholders, including the customers, employees, investors, suppliers, and the larger communities in which the business participates."[19]This opened my eyes to a new way of thinking about companies. As an analyst, I had always thought of a company's stakeholders as individual risk factors (i.e., customer trend risk, labor risk, dependence on suppliers, environmental risk), but never in the holistic sense. What I loved was that Mackey's concept of stakeholder capitalism tied perfectly into my thesis of social media being an exposing disruptive force as this new medium empowered a company's stakeholders by giving them a voice.

This ideology resonated with me, appealing as it did to my positive mindset. Unlike the world of social responsible investing, which at the time was more focused on looking at a company's ESG (environmental, societal, governance) factors from a negative screening perspective based on the principle of fiduciary duty, the "conscious capitalism" world was focused on promoting the more positive ideology that companies should think about the best interests of all their stakeholders—not just their shareholders—and most importantly, make profit with a purpose.

As I started to dive into the ideology of companies having a greater purpose, I came across these wise words of Fred Lager, former CEO of Ben & Jerry's, in his 1994 book *The Inside Scoop:*

How Two Real Guys Built a Business with a Social Conscience and a Sense of Humor:[20]

> *The issues here are heart, soul, love and spirituality. Corporations that exist solely to maximize profit become disconnected from their soul—the spiritual interconnectedness of humanity. Like individuals, businesses can conduct themselves with the knowledge that the hearts, souls and spirits of all people are interconnected; so that as we help others, we cannot help helping ourselves.*

This kind of thinking was by no means common at the time Lager wrote these words, nor was it in January 2011 when I made my recommendation for investors to buy lululemon and sell Yellow Media. In fact, it was viewed as a pretty risky call. From a quantitative perspective, Yellow Media, at C$6.26, appeared to be an attractive buying opportunity—it was trading at only 7.8 times earnings and offered an appealing 10% dividend yield. At the same time, lululemon, at US$67.20, was trading at an astronomical 48 times earnings and appeared as a great potential short candidate. But in the months after publishing my first report, as I watched the stock price of "heart and soul" lululemon rise while the stock price of "shell" Yellow Media fell, I became increasingly confident that I had discovered a new way to look at and value stocks.

And my investment thesis continued to play out beyond my wildest expectations. By the time I published my second report, the stock price of lululemon had risen by over 66% to US$55.98 (it split two for one on July 12, 2011) while the stock price of Yellow Media had fallen by 94% to C$0.33. And on August 3, 2011, Yellow Media's credit rating was downgraded from BBB- to BB+ by Standard & Poor's, and the company cut its dividend from C$0.65 to C$0.15. Less than two months later, the company took a C$2.9 billion goodwill impairment charge and announced that the dividend would be completely eliminated.

Eager to put my newfound investment thesis to work for my own investment portfolio, in February 2011 I put together an initial list of publicly traded heart and soul companies that included Apple, lululemon, Whole Foods Market, Starbucks, Costco, Nordstrom, Build-A-Bear Workshop, Amazon, Nike, and Netflix. In narrowing down my list, I unfortunately missed the boat on Apple, Amazon, and Netflix, as I was more confident in my ability to analyze consumer stocks than tech stocks. After doing due diligence on the remaining consumer companies (reading through each company's annual reports, quarterly reports, and quarterly conference call transcripts), I settled on lululemon, Starbucks, and Whole Foods Market. The fourth and final company I selected was Chipotle Mexican Grill, a company I discovered through Jim Cramer, whose CNBC show *Mad Money* I was (and still am) a devoted fan of.

Ironically, even though I had been researching and writing about social media, I was still stuck in somewhat of a scarcity mindset, as I had never experimented with having my own website or with blogging or tweeting. But on August 24, 2011, I decided to take the plunge into the world of micro-blogging and created my first Twitter handle (@barbcfa). And in November 2011, just a week before the publication of my second research report, I launched the Brady Capital Research website.

I came up with the concept for Brady Capital Research while on my honeymoon—a five-week road trip (with nine-month-old Brady!) down the California coast to my parents' vacation home in Palm Springs. When we arrived back home in early November, Greg created the website. I featured Brady's picture on the homepage, along with the tagline "Brady Capital Research—For the Next Generation of Investors." I also created the BCR Library to feature my recommended business strategy books, classifying them into five categories: business strategy, founders, innovation, leading edge, and greater purpose. To add more color, I included what I felt were the top three quotations from each book, and made it all socially interactive by adding the LinkedIn, Twitter, and blog feeds for each author.

By researching companies that were making a positive difference in the world, starting with my initial list of four companies, I hoped to come up with new ways to look at and value companies, as well as to create a community that brought together heart and soul companies with investors. While I achieved my first objective, I failed miserably at the second, ironically failing to recognize that there was a scarcity of supply of publicly traded heart and soul companies, and also a scarcity of demand for my type of qualitative-oriented research—the majority of investors didn't care about how companies treated their stakeholders or if they had a social mission.

It seems now that I was ahead of the curve in terms of recognizing the power of advocacy as a social value driver: in recent years, we have seen an increasing number of companies and investment firms hopping on the "social mission" and "greater purpose" bandwagons. In fact, in June 2016, Perdue Farms, one of the largest producers of poultry in the United States, announced a plan to focus on chickens' wants and needs. This is a testament to the power of advocacy (thanks to the democratization of influence through social media, which has given everyone a voice—apparently chickens too) as a new social value driver.

To Create Advocacy, Start with the Four Cs

In November 2011, I published the thematic research report *Social Capital: A New Strategic Play for Investors; Look for Companies with Heart and Soul.* My basic thesis was that by creating positive social capital (shared values and positive externalities) for stakeholders, companies would be able to create a thriving stakeholder ecosystem. As Dev Patnaik states in *Wired to Care*, "When companies blur the line between us and them, between inside and out, opportunities appear for mutual benefit. Customers who become loyal fans energize the brand. Suppliers become partners who bring new ideas to the table."[21]

In the report, I recommended Chipotle Mexican Grill, lululemon, Starbucks, and Whole Foods Market as companies with heart and soul, as they met my Four Cs criteria:

- **Conception.** Greater purpose and unique DNA create a wide economic moat.
- **Core values.** Authentic core values serve to guide the company.
- **Community.** Cult-like following creates a strong stakeholder foundation.
- **Culture.** Organic growth strategy preserves cultural DNA.

Conception: Greater Purpose and Unique DNA Create a Wide Economic Moat

My research made it clear that one of the important differentiators of a heart and soul company is its conception. The founder plants the seeds for the movement because he or she is inspired by a greater purpose and conceived the company through passion. Therefore, because the heart and soul company is imprinted with its founder's DNA, "nature" is a key influencer.

This was apparent in the four companies I focused on in my research. In 1978, John Mackey planted the seeds for his Whole Foods Market movement, inspired by the mission to "help support the health, well-being, and healing of other people and the planet." In 1987, Howard Schultz started his Starbucks movement to "inspire and nurture the human spirit—one person, one cup and one neighborhood at a time" by buying the six-unit coffee shop chain from its founders. In 1993, Steve Ells began Chipotle Mexican Grill, inspired by the simple philosophy to demonstrate that food served fast doesn't have to be a traditional fast-food experience, and this vision has evolved to now change the way people think about and eat fast food, guided by Ells's mission of "food with integrity." In 1998, Chip Wilson, inspired to "create components for people to live a longer, healthier, and more fun life," started his lululemon movement.

With all of this in mind, I proposed in my report that, although competitors may attempt to nurture their companies by copying companies like lululemon or Chipotle Mexican Grill, their efforts would not be successful if they were missing the nature element: the original DNA of the founder. The unique DNA and greater purpose of a heart and soul company would create an enduring, nonreplicable, intangible asset, which would result in a wider economic moat. However, what time has proven since my report was published is that, although the original DNA is a valuable intangible asset, the imprint can start to fade if a founder decides to leave or to become less involved in the company. In retrospect, this explains Howard Schultz's decision in 2008 to return to Starbucks from his eight-year sabbatical when the company was in need of having its brand and purpose restored.

Core Values: Authentic Core Values Serve to Guide the Company

Although all companies publish and tout their core values, many of the core values they cite are generic and inauthentic, not truly reflective of the reality of what a company does and how it actually operates and treats its various stakeholders. However, a differentiating element I discovered in the values statements for the heart and soul companies is how original and descriptive these companies were in articulating their core values, and how many of the core values were connected to the greater purpose, or soul, of the companies and spoke to the interests of all the stakeholders.

Whole Foods Market cites as one of its core values "promoting the health of our stakeholders through healthy eating education." In Starbucks' list of principles of how it lives its mission every day, it encompasses its coffee, partners, customers, stores, neighborhoods, and shareholders with phrases such as "It's about the enjoyment at the speed of life—sometimes slow and savored, sometimes faster. Always full of humanity" and "We can be a force for positive action—bringing together our partners, customers, and the community to contribute every day."

Chipotle Mexican Grill advocates its "food with integrity" mission by addressing the topics of animals, people, and the environment, saying, "We work hard to find farmers and ranchers who are doing things the right way," and "We can talk about all of the procedures and protocols we follow and how important they are—but it really comes back to the people," and "We work hard to source ingredients in ways that protect this little planet of ours."

Interestingly, three of the seven core values that lululemon espouses include inspirational values as such:

Fun—*When I die, I want to die like my grandmother who died peacefully in her sleep. Not screaming like all the passengers in her car [a riff on a Will Rogers quotation].*

Balance—*There is no separation between health, family and work. You love every minute of your life.*
Greatness—*We create the possibility of greatness in people because it makes us great. Mediocrity undermines greatness.*

Community: Cult-Like Following Creates a Strong Stakeholder Foundation

The greater purpose and authentic core values of these heart and soul companies attract a cult-like following from people who believe in the movement's greater purpose and are seeking to align their values with those whom they buy from, work for, and do business with. As the followers are inspired by their heart and soul to join the movement, this results in the companies building an authentic and strong stakeholder foundation—a community—in terms of:

- **Employees.** Heart and soul companies are able to attract high-caliber, passionate, and motivated candidates who seek intrinsic satisfaction by working for companies whose corporate values align with their personal values and provide them with a sense of community, purpose, and meaning. As C. William Pollard writes in *The Soul of the Firm*, "People want to work for a cause, not just for a living."[22]
- **Customers.** By virtue of their greater purpose and authentic core values, heart and soul companies are able to attract consumers to their product or service through local grass-roots marketing initiatives, instead of having to resort to

traditional advertising and marketing campaigns or celebrity endorsements.

- **Business partners.** Heart and soul companies also attract business partners to their movements that seek to work alongside companies whose values resonate with their own and who they can join forces with to help make a positive difference in the world. These business partners range from suppliers to communities to investors.

Culture: Organic Growth Strategy Preserves Cultural DNA

My research led me to realize that, in order to preserve their cultural DNA, most heart and soul companies made a conscious strategic decision to maintain control over their operations. They achieved this by pursuing a business model of corporate-owned stores as opposed to franchising. In addition, they elected to grow mainly through organic means rather than by acquisition.

Although Starbucks had never grown through acquisition or franchised its operations, it strategically decided to enter into licensing agreements in order to gain access to otherwise unavailable real estate such as airport locations, national grocery chains, major food services corporations, and institutional foundations like college campuses and hospitals. The Starbucks movement was in full force from 1987 to 2008—during this 21-year period, Howard Schultz expanded his coffee shop empire to nearly 17,000 stores, over 9,200 of them company-operated. However, even the seemingly bulletproof company wasn't immune to the effects of the Great Recession: it brought the expansion to a halt, and the company closed 800 Starbucks stores in the United States and 100 internationally.

Chipotle Mexican Grill was the purest of my four heart and soul companies in terms of organic growth. Unlike most restaurants, the company continued to expand during the recession. When I spoke with management about its corporate philosophy on franchising and acquisitions, I was told the company prefers to grow within so that it can control the people culture, thereby preventing it from ending up with people who do not fit and supplier relationships that do not work.

Of the four heart and soul companies I honed in on, lululemon is the youngest, founded as it was in 1998. The company distributes its athletic apparel and accessories exclusively through its network of corporate-owned stores. The company decided to close its franchise program back in 2007 and opportunistically bought back its franchises. The higher business risk involved with franchisees was laid out by lululemon in its 2010 annual report:[23]

> *We do not exercise control over the day-to-day operations of their retail stores... franchisees may not successfully operate stores in a manner consistent with our standards and requirements, or may not hire and train qualified employees, which could harm their sales and as a result harm our results of operations or cause our brand image to suffer.*

Whole Foods Market operates natural and organic food supermarkets, the majority being in the United States, with just a handful in Canada and the United Kingdom. Although Whole Foods Market operates its own stores and does not franchise, the company has pursued growth through acquisition. The company made its largest acquisition in August 2007 when it acquired Wild Oats, a North American organic grocery store competitor with 110 locations, for US$565 million. However, it was forced by the US Federal Trade Commission on antitrust grounds to sell the chain in March 2009, and it retained only 7 operating and 17 nonoperating stores. In my conversation with management, the company said it had no plans for franchising, as it wanted to preserve the brand and had the capital to grow. In terms of acquisitions, it would still do small acquisitions but did not see a lot of opportunity, given its non-union status. The company would do real estate acquisitions to pick up the leases and might look to expand internationally through acquisitions.

As of May 2016, four and a half years after my report was published, all four of these heart and soul companies have expanded their geographic footprints while staying true to their organic growth strategies. And the unit growth for all four is impressive. As of the spring of 2016, lululemon's base is up 147% (from 147 to 363 stores), Chipotle Mexican Grill's is up 78% (from 1,163 to 2,066 restaurants), Whole Foods Market's is up by 43% (from 311 to 446 stores), and Starbucks' is up by 40% (from 17,000 to nearly 24,000 locations).

The thesis I presented in my report was that, by building their employee, customer, and business partner bases from the ground up and not taking any growth shortcuts through acquisitions or franchising, heart and soul companies would have stakeholders who were likely to be engaged and in line with the companies' core values. This would provide the companies with greater control and influence, and further strengthen their stakeholder foundations.

***

To access the new social value driver of advocacy, companies need to pay attention to all of their Four Cs: conception, core values community, and culture. To summarize: First, a company needs to stay true to its conception—the founder's unique DNA and inspiration for starting the company. Second, a company needs to identify its authentic core values. Third, a company needs to figure out how to build a thriving ecosystem by identifying all the stakeholders in its community. And fourth, a company needs to make sure it has a strong corporate culture that is aligned with the company's purpose and core values.

Over the long term, following this strategy will also enable a company to create a higher level of trust and commitment on the part of its stakeholders and will result in higher employee productivity and lower turnover, higher customer loyalty, greater share of wallet, higher pricing power, and greater cooperation from business partners. I won't get into the details, but in my *Social Capital Play* report I also discuss in depth how advocacy would accelerate the value creation process for heart and soul companies by lowering their risk profile and acting as a growth catalyst to grow comparable store sales, expand their geographic footprint, and launch new concepts.

5

How Rebels with a Cause Build Advocacy

The First Step to Becoming a Rebel: Finding Your Voice

As an analyst, I had been trained to write research from an impersonal and soulless perspective and so, as you might expect, blogging wasn't something that initially interested me. It didn't help that I was of the (erroneous) belief that writing a blog wasn't something professionals did, and I didn't want to be labeled a "mommy blogger." But thankfully, a discussion I had with Carol Sanford changed my mind.

Carol is the author of *The Responsible Business*, a book which, like many I was reading at the time, focuses on the stakeholder concept. Of all the mentors I encountered on my journey, I consider Carol to be one of the most influential, as she encouraged me to start blogging in order to share my views in a real-time manner, and she recommended I write in less of a business-like tone. This was a big departure for me, and a challenge to weave together analysis and narrative. But once I tried blogging, I realized I'd found what I never knew had been missing all these years—my own voice.

I conquered my fears and published my first blog article on

March 17, 2012, which I titled "Note to Trump and Buick: A Brand Creates Equity Through What It Does, Not What It Says..." I find it ironic that as I write this four years later, Donald Trump is the Republican candidate for the American presidency. But looking back, what I find most fascinating about my first blog article is that it provides a window into my soul and the awakening of conscience that was starting to take place. For example, I had written: "I have to admit I aspired to live the 'money, money' lifestyle and ride around in private jets and limos. But over the past two years, I have had somewhat of an awakening of conscience and have realized that the world I have been living in is starting to come crashing down."

Note to Trump and Buick: A Brand Creates Equity through What It Does, Not What It Says...

March 17, 2012—Last night after we got our baby to sleep, my husband, Greg, and I curled up on the couch to indulge in one of our favorite reality TV shows, *Celebrity Apprentice*. It is interesting, as this season I have become increasingly disillusioned with Trump, and I have been questioning his command-and-control style and his overriding "it's just business" mentality. But it wasn't until we watched last night's episode that I realized how out of touch corporate America is becoming. The basic premise of *Celebrity Apprentice* is that you have two teams made up of celebrities who compete with each other every week to raise money for their chosen charity. The assignment this time was to create a presentation for Buick's new Verano car; the leader of the winning team would win US$20,000 plus an additional US$30,000 donated by Buick for their charity.

I started to get a bit suspicious when Trump started to question Michael Andretti for not stepping up to the plate as a team leader.

My skepticism grew during the first commercial break, which seemed to be sponsored by guess which company? Yep, Buick—a definite product tie-in. But it got even worse, as not only did the men's team (which, in my opinion, did a much better job with their presentation) lose, but Trump ended up firing both the team leader (Adam Carolla) and, in a very contrived way, Michael Andretti, who did absolutely nothing wrong. My gut told me that Trump must have pitched Buick on the promise of an endorsement from racing legend Michael Andretti, and when Andretti did not provide this, Trump decided to fire him.

The whole thing was particularly interesting to my husband and me because we had just returned from the SXSW Interactive Conference in Austin, Texas, where Chevy (a sister to Buick in the GM family) and GE (parent to NBC, which broadcasts *Celebrity Apprentice*) were both sponsors. And the message we heard over and over again at the conference was how the old push-style method of advertising no longer works, and how companies have to be transparent and authentic and engage in a real discussion with their customers. I find the total lack of transparency and inauthentic behavior by both Trump and Buick appalling, and their thinking they got away with trying to manipulate Andretti is insulting to their viewers and customers. I guess nobody at corporate GM or GE headquarters was sitting in on any of the SXSW presentations with us.

But thanks to social media, these shells are being exposed. My sense is that this debacle will create negative social capital for both Trump and Buick. Although it will take a while for the numbers to play out, I believe *Celebrity Apprentice* has really lost trust with celebrities, advertisers, and its viewers, which will impact the future value of its franchise. And Buick not only got a negative ROI from this venture, but it seems to have tarnished the brand equity of its new Verano vehicle. As Greg Johnson, global creative director at HP, remarked during the "Brands as Patterns" presentation at the SXSW conference, "A brand creates equity through what it does, not what it says."

***

Now, four years later, the sun is indeed setting on the Verano (Buick plans to discontinue it in the fall of 2016), though I was clearly wrong about Trump. His sun is shining brighter than ever: there is a chance he could be the next president of the United States. (Ironically, Arnold Schwarzenegger, former governor of California, is taking over as the host of *Celebrity Apprentice*.)

To achieve success in the ubernomics era, companies will need to create a movement, not just a marketplace. And, as I explore further in Chapter 9, creating a movement starts with being a "rebel with a cause" and relishing the disruption of the status quo. For this reason, I feel it's also worth sharing the articles I wrote in 2012 that explore how rebels with a cause (namely Howard Schultz, Christine Day, and Steve Ells) create advocacy. Let's start with my blog about Schultz and how he's creating advocacy at Starbucks. But first, some context.

The week after I wrote the blog article about Trump and Buick, I drove down to Seattle with Greg and baby Brady to meet up with Norman Wolfe, the author of *The Living Organization*. After reading his book the previous year, I had connected with Norman through the Conscious Capitalism group on LinkedIn. I found the book quite fascinating; in it, Norman looks at companies from the perspective of quantum physics and espouses radical views such as: "Our businesses are living beings. They are not soulless machines that are only concerned with maximizing production and shareholder value. Yet that is what they have become. They have honed their ability to produce but lost their ability to contribute."[24]After my meeting with Norman at Starbucks' Roy Street Coffee & Tea on Capitol Hill in Seattle, Greg and I decided to stop in Bellevue, Washington, on our way home, to check out Starbucks' new Evolution Fresh concept store. This experience is what inspired me to write my next article.

Evolution Fresh: An Exciting New Chapter in Starbucks' Movement to "Inspire and Nurture the Human Spirit"

March 22, 2012—When traveling back from Seattle recently with my husband, we took the opportunity to check out Starbucks' new Evolution Fresh concept store. What struck me first was the great ambience, high positive energy, and organic feel of the place. The Smooth Blue was delicious (one of the better smoothies I have tasted!), and the quinoa, kale, and butternut squash dish was amazing—it tasted incredibly fresh and healthy—such a great combination of flavors and good-for-you ingredients. And the rest of the items on the menu look equally yummy and healthy.

I think Starbucks' entry into the US$50 billion health and wellness market through the launch of its new Evolution Fresh concept underlines how the higher empathy level of heart and soul companies allows them to recognize opportunities faster than their competitors. If you think about it, Starbucks has already won the hearts and minds of its customers (60 million people visit Starbucks every week!) and will be able to leverage the same employee culture, grassroots marketing strategy, and existing infrastructure of production sourcing, logistics, and management information systems. Worth noting is that the couple sitting next to us (I heard the girl exclaim, "This is my new favorite place!") did not know that Starbucks was behind Evolution Fresh, as there was no obvious Starbucks branding aside from the fact Starbucks coffee was on offer. But in talking with the employees, we discovered that half came over from the Starbucks next door and that Evolution Fresh is piggybacking off the store's free WiFi connection (which I love!) and using Starbucks' customer-card platform. There will be cross-marketing opportunities too, as Starbucks' plan is to replace the bottles of Naked Juice it currently buys from PepsiCo with its own Evolution Fresh lineup, starting in its Washington stores this spring. Interestingly, the guy who Starbucks bought Evolution Fresh from back in November for US$30 million was Jimmy Rosenberg, who also founded Naked Juice, which he sold to PepsiCo for an undisclosed sum back in 2006.

I am really excited to see how Starbucks will use social media to leverage its existing strong relationships with its employees, customers, business partners, and communities to introduce and launch Evolution Fresh. The company's success in using social media to launch new concepts is evidenced by its successful launch of the VIA instant coffee product back in October 2009—when it had only 5 million fans on Facebook and 450,000 followers on Twitter—to reach sales of US$100 million in only 10 months. Starbucks' current extensive social media reach of 40 million fans on Facebook, 2 million followers on Twitter, and nearly 100,000 followers on LinkedIn will prove to be a valuable asset to raise awareness for its new concept and draw patrons into Evolution Fresh. The conversational style of Twitter seems to make it the platform of choice for Evolution Fresh: the company's newly created @EvolutionFresh Twitter account already has 605 followers, but it has not yet launched its Facebook page (though there is a community page with 1,494 likes) nor a separate LinkedIn company account.

Starbucks' management will be able to use the customer feedback to really listen and engage with its customers, which will allow it to make adjustments and improvements to the Evolution Fresh pilot store before it invests future time and capital to expand the concept. The company is engaged in grassroots listening—there was a table of health and wellness bloggers sitting next to us, providing their feedback and views on the food and the store itself, and one of Starbucks' culinary experts who had created the dishes was circulating about the store, giving out samples (so tasty!) and actively engaging with customers.

I believe Evolution Fresh will be able to capitalize on the high level of social capital that Starbucks has built over the past 25 years through its authentic core values, caring culture, and social conscience. More importantly, Evolution Fresh will provide the company with a platform to raise awareness and educate its customers about nutrition and the benefits of healthy eating I believe Evolution Fresh represents an exciting new chapter in the movement Howard Schultz started back in 1987 to "inspire and nurture the human spirit—one person, one cup and one neighborhood at a time."

***

Although Evolution Fresh never took off as a standalone store concept, Starbucks has been successful in leveraging the health-oriented brand through its network by selling Evolution Fresh juice in 11,000 of its Starbucks stores, offering smoothies made of Evolution Fresh juices, and partnering with Danone to launch a line of Evolution Fresh Greek yogurt.

A few days after my blog post about Evolution Fresh, lululemon reported its quarterly earnings. My apprehensions about blog writing now completely gone, I was inspired to write another article. As lululemon's stock price had risen by 35% in the five months since I had recommended it in my research report *Social Capital: A Strategic Play for Investors,* I was a bit nervous that the company would not meet investors' heightened expectations.

Social Capital: The Secret to lululemon's Success

March 24, 2012—I woke up at 5:24 a.m. yesterday and reached over to the nightstand for my iPhone (I ditched my BlackBerry back in January 2010, when I elected to leave the corporate world) and immediately went to the CNBC app to check out the stock price of lululemon, which was set to report its 4Q11 results before market open. I breathed a sigh of relief when I saw that the stock was trading down only 2%. Although I'm a strong believer in the company's long-term growth prospects, and I view it as a "baby Nike," its forward price-to-earnings (P/E) multiple of nearly 50 times makes it an ideal short target for investors who are just looking at the numbers.

Although I no longer officially cover lululemon, I have a strong personal affinity to both the stock and the company. In addition to being one of the first analysts to initiate coverage on it shortly after its IPO in September 2007, I used to live at 3rd and Arbutus, just around the corner from lululemon's original store in Kitsilano, Vancouver. And as a professional female who lives an active and healthy lifestyle, I am its picture-perfect target customer. Although I am not a huge yogi (I prefer more active pursuits, since I can never seem to get my type-A mind to sit still), I am a huge fan of lululemon and I have the wardrobe to prove it!

Anyway, back to the financials... I took a quick scan through the company's press release—another impressive quarter! Revenue was up 51%, with comparable store sales growth of 26%—and that was on top of 28% growth a year ago. Earnings per share (EPS) came in at US$0.51, up from US$0.38 last year—a 38% gain. And the company ended the year with US$409 million in cash and equivalents (gotta love a company with no debt!) and 174 stores.

I tiptoed into the living room (hoping Brady would not wake up) to log onto my computer in order to listen to the company's analyst conference call. To be honest, most company conference calls put me to sleep, since management tends to just focus on the numbers. But the conference calls of companies with heart and soul, like lululemon, are different—the tone is much more qualitative, and you can tell that management actually cares about its stakeholders—customers, employees, suppliers, community, the environment. This was evident in the lululemon call when Christine Day, president and CEO, started off by sharing the news that lululemon had just reached the US$1 billion annual revenue milestone and commenting:

> *But far more important than the number itself are the beliefs, values, culture, and people that achieved it, and our guests who value our products and our guest experience. Our success now and in the future is based on a culture of high performance and leadership development. We find the right talent, empower our employees, teach personal accountability and judgment, and share our business strategy at every level across the business to ensure that we engage our employees and give them a sense of purpose.*

I believe the strong and authentic stakeholder foundation of a heart and soul company like lululemon will convert into a high velocity of social capital as the company leverages the high level of enthusiasm and deep psychological attachment to the company's brand and great purpose. And after sitting through five days of presentations at the recent SXSW Interactive Conference in Austin, Texas, I am even more convinced in my social capital investment thesis, since the message I heard over and over again was how companies need to be authentic, transparent, caring, and actually engage with their customers. How else can you explain how lululemon, a company that sells premium-priced discretionary goods, can attract customers to its stores and achieve astronomical sales per square foot of over US$2,000 and a gross margin of over 55%?

There is no doubt lululemon will continue to have its skeptics and be a favorite of shorts—most analysts and investors see no need to look beyond the numbers. Only time will tell if my social capital investment thesis proves correct. And in the meantime, with spring finally here, I can't wait to check out lululemon's new line of cycling apparel (it will be nice to actually look good when I cycle!), and I am excited to see the upcoming release of its new swim and surf collection.

***

Around this time, Starbucks held its annual general meeting. I wished I could have gone down to Seattle to attend it in person, but logistics were too difficult with Brady, so I had to make due with logging into the webcast. I have to admit, this was the first time I attended (albeit virtually) a Starbucks meeting, and wow, what a difference it was from your typical boring, soulless, corporate one. I was so moved by it and the words spoken by Howard Schultz that I felt compelled to share my thoughts in another blog article, which you'll find on the next page.

Although I have not yet had the pleasure of meeting Mr. Schultz, I have a deep admiration for him. Looking back to 2006 when I read his book *Pour Your Heart into It: How Starbucks Built a Company One Cup at a Time,* I realize now that my starting to think about companies in a more holistic sense was really the turning point in my career as an analyst. And part of my original conviction about lululemon came from the parallels I saw between it and Starbucks. As Schultz shares in his book, "Our goal was to build a great company—one that stood for something—one that valued the authenticity of its product and the power of its passion."[25] Fittingly then, there was a real sense of passion and caring conveyed throughout the meeting I listened in on that day. It lasted over two hours, and I took 10 pages of notes, the highlights of which I included in my blog post.

Starbucks: The Value in Values

March 26, 2012—I listened in to Starbucks' annual general meeting (AGM) a few days ago via the webcast, and I couldn't help but be impressed. I loved how Schultz started his speech talking about the values that define the company and how Starbucks is a different kind of company, as it has been built through the lens of humanity. And near the end of the meeting, the company's CFO Troy Alstead, emphasized that "our value comes from our values" and talked about how important it is to do the right things for the right reasons, and related how Starbucks' shareholder value is derived from its core values. Although this makes intuitive sense unfortunately, this is a radical departure from the views of most investors, who suffer from financial myopia. But to these cynics I say take a look at what Starbucks has achieved since its IPO.

When Starbucks went public in 1992, the company had 300 employees, 150 coffee houses, and less than US$150 million in revenue. Over the past 20 years it has grown into a global coffee empire, with 149,000 employees, 17,000 coffee houses in 58 countries, and revenue of US$12 billion. And how did the shareholders do? Well, Starbucks' market cap was US$250 million when it went public and just reached US$40 billion—up 160-fold—which by my math equates to a compound annual growth rate (CAGR) of 29% a year!

I am ashamed to admit that when I used to research companies, it did not even occur to me to look at a company's core values. And looking back to when I studied for the Chartered Financial Analyst (CFA) designation, even though the curriculum places a huge emphasis on ethics, I do not recall any of the textbooks covering the topic of core values. I believe this is a significant oversight. As I do more and more research, I realize how important it is for companies to have an authentic set of core values to guide them. Starbucks is a great example of this, as one of its principles of how it lives its mission is "We can be a force for positive action—bringing together our partners, customers, and the community to contribute every day." And this is not just talk.

During his speech, Schultz shared that Starbucks will be giving the profits back to the community for two of its stores in LA and Harlem. In addition, at a time when most companies are electing to reduce costs by locating new plants outside the United States, Starbucks has decided to open a roasting facility for its VIA instant coffee product in Augusta, Georgia, which will create 150 to 200 new jobs. And on top of this, Schultz decided to take the unorthodox approach of using Starbucks' scale for good and created a jobs initiative program, which has raised millions of dollars from over 600,000 customers donating money to establish a national network of loans for small businesses. The AGM actually took an evangelical tone as Reverend Dr. Calvin O. Butts III, pastor of the Abyssinian Baptist Church in Harlem, took the stage and shared his admiration for Schultz, noting that what he represents is so important.

And Starbucks is not just sitting still. During the AGM, Jeff Hansberry, president of Channel Development and Emerging Brands, spoke about how the company will be able to leverage the deep emotional connection consumers have to the Starbucks brand with its unique ecosystem of Starbucks stores, retail distribution channels, and strong digital presence (40 million Facebook fans and 2 million Twitter followers) to launch innovative new concepts. At the AGM, Starbucks announced the launch of three new concepts: (1) the Verismo system (at-home, premium, single-cup espresso machines it developed with Germany's Krüger GmbH & Co.), (2) Starbucks Refreshers (a new natural energy product that leverages new technology for green coffee extract), and (3) Evolution Fresh (its new fresh juice/health store concept). I believe the most important new concept from a strategic perspective is Evolution Fresh, as it marks Starbucks' entry into the US$50 billion health and wellness market.

Cliff Burrows, president of Americas, spoke about how the Starbucks card is now used for 1 in 4 transactions and how there were 26 million mobile transactions last year. He also mentioned how Starbucks is launching a crowdsourced community program called Vote.Give.Grow, whereby customers can vote on how $4 million is allocated to nonprofits. Michelle Gass, president of Europe, Middle East, and Africa, who architected the US turnaround, spoke about how she is working to reintroduce Starbucks to Europe by retraining all the baristas to create personal connections with their customers by putting names on the customers' cups. John Culver, president of Asia and China, shared how Starbucks formed a $1 million cup fund to support its partners in Japan who were impacted by the earthquake, and he spoke about how values truly do transcend cultures. He also spoke about how Starbucks plans to build its brand in a holistic way and have over 1,500 stores in mainland China by 2015, representing Starbucks' highest retail growth opportunity, with the added benefit of the region offering the highest profit margins.

The bottom line is that the strong and authentic stakeholder foundation of a heart and soul company, like Starbucks, will convert into a high velocity of social capital. This will accelerate its value creation process by serving to lower the company's risk profile and act as a growth catalyst for it to continue to grow its comparable store sales, further expand its global footprint, and launch new concepts (e.g., Evolution Fresh). And throughout Starbucks' AGM, my mind kept wandering back to the inspiring speech delivered by Tim O'Reilly, founder of O'Reilly Media, at the recent SXSW Interactive Conference—about how companies need to be held to a higher standard—and his advice of not just doing "do-goodism" but doing good by "doing business that matters."

***

I love how Howard Schultz has continued to build advocacy for Starbucks by being a rebel with a cause and challenging the status quo. For example, in June 2014, he launched the Starbucks College Achievement Plan in partnership with Arizona State University to provide all its full-time and part-time US employees the opportunity to receive reimbursement on college tuition for their junior and senior years. And in April 2015, Starbucks expanded the plan to cover the entire cost for its employees of getting an online degree.

Another rebel-with-a-cause founder I admire who has built advocacy through caring for his employees and creating a unique people culture is Steve Ells, founder of Chipotle Mexican Grill, which I wrote about in this blog article.

Chipotle: A Story Worth Sharing

April 20, 2012—I was a bit nervous on April 19 after market close as I waited for Chipotle Mexican Grill to report its 1Q12 results. Although I strongly believe in the company's long-term value creation potential, the stock had risen 16% (versus only a 3% gain in the market) in the past quarter and was trading at a premium valuation of nearly 40 times next year's earnings. I breathed a sigh of relief as the headline numbers scrolled across my screen.

I have to admit, I am a latecomer to the Chipotle story. I had always been skeptical because I dislike fast food, and I had never visited a Chipotle, as it had only two restaurants in Canada (in Toronto, not Vancouver). And so I didn't feel the need to look into the company until one afternoon while watching *Mad Money* (I think poor Brady watches more hours of CNBC than cartoons!) and host Jim Cramer started singing the praises of Whole Foods Market—and then Chipotle Mexican Grill. I decided it was time to put on my analyst hat and take a closer look.

To start, I checked out the company's website. I really liked the playful and creative tone—it felt authentic and human. And as I started to dig into the story behind Chipotle, I discovered that its founder, Steve Ells, is a graduate of the esteemed CIA (Culinary Institute of America), and that the company is about much more than just selling burritos. Chipotle advocates its "food with integrity" mission by addressing the topics of animals, people, and the environment. I also respected that the company built its entire restaurant base from scratch, resisting the temptation to accelerate its growth through taking acquisition or franchising shortcuts. And as I read through the past few years of quarterly conference call transcripts, I was struck by how genuinely passionate management was about its "food with integrity" mission.

Excited that I might have found a company with heart and soul, I decided to crunch the numbers and figure out Chipotle's business model, so I pulled up its annual and quarterly reports for the past few years. The numbers looked impressive: in 2010, Chipotle earned $1.8 billion in sales from its 1,020 restaurants, equating to average sales per restaurant of $1.8 million. And through the dark years of 2008 and 2009, the company managed to achieve both positive same-store sales growth and double-digit unit growth—likely a testament to its strong stakeholder equity foundation. But it wasn't until last spring, when I drove across the border to Bellevue, Washington, and walked into the restaurant (the closest location to Vancouver) that I understood why people fall in love with Chipotle and become loyal and frequent guests. Even though it was 2 p.m. on a rainy Sunday, there was a long line-up at the counter. I was impressed with how friendly, professional, and efficient the front-line crew was. But what blew me away was the food itself—so fresh and flavorful. I swear, their guacamole is the best I have ever tasted. I was sold!

During the analyst conference call, Steve Ells talked about how the major drivers of the company's business are its unique food and people culture—and this theme continued throughout the call. I have to point out that this is not the norm. Most conference calls are focused on just the numbers and how the company is achieving operational and cost efficiencies. One of the highlights of the quarter was the increased awareness and public dialogue Chipotle generated for its "food with integrity" greater purpose when it aired its two-minute animated *Back to the Start* film during the Grammys. According to management, this multi-award-winning, soul-stirring film—set to Coldplay's song "The Scientist," sung by Willie Nelson—generated 22,000 tweets, over 33 million Twitter impressions, and some 150 million media impressions. This really highlights how authentic companies with a greater purpose are able to leverage social media to build emotional connections with customers and psychological attachments to what the company stands for.

The bottom line is that companies with heart and soul, like Chipotle, are best positioned to outperform in this decade of the social revolution. I am convinced that companies that derive their competitive advantage by exploiting their stakeholders will start to see their economic moats erode. Meanwhile, heart and soul companies, like Chipotle, that are working to build higher levels of social capital will be able to create new generative moats and achieve higher-than-expected growth. And that is why I believe Chipotle is a story worth sharing.

***

Four years later, I am questioning whether Chipotle is still a story worth sharing. Although I also loved the company's second short film (the haunting but beautiful *The Scarecrow*, which has been viewed over 16 million times on YouTube since being released in September 2013), the long-awaited follow-up (*A Love Story* released in July 2016) doesn't resonate with me. The film ends with the same inspiring message to "cultivate a better world," but the love-story theme doesn't ring true. The film is lacking in authenticity and even seems a bit vindictive. I love how Chipotle's first two films advanced the company's mission to "change the way the world thinks of and eats fast food" by seeking to raise awareness and educate people about how their food is produced and processed. But *A Love Story* is rather dark by contrast and comes off as a much more direct attack on the marketing tactics of its fast-food competitors. This seems ironic given that Chipotle has strayed from its grassroots marketing approach by engaging in a heavy promotional campaign to win back customers who were lost in the six months following the company's food-poisoning scandal. I wish it weren't the case, but between this and the recent cocaine bust of one of its senior executives, it seems like Chipotle may be losing some of its heart and soul.

Investing in companies led by rebels with a cause is not for the faint of heart. It's important to remember that things that go up can also come down. And in the two months after I wrote the Chipotle article, that's exactly what happened: my high-multiple growth stocks came crashing down with the market, as my next blog article attests.

lululemon: Rebel with a Cause

June 11, 2012—lululemon is rebelling against Wall Street's myopia. Why does the company believe in investing precious time and capital to deepen relationships with its customers, employees, business partners, and communities? And how does it have the audacity to consciously sacrifice short-term profits for potential long-term gains? Is not the obligation of a corporation solely to maximize profit for its shareholders?

Given that we are still in a weak economic environment and lululemon sells premium-priced apparel, you would expect investors to be joyous over a company that is able to grow its top line by 53% (with comparable sales up 25%—on top of 16% growth a year ago) and its bottom line by 39% (EPS came in at US$0.32, up from US$0.23). But instead, investors took the stock down by 10%, as they started to panic over the 370 basis-point fall in its gross margin and were disappointed the company maintained its guidance for the year of revenue of US$1.32 to US$1.34 billion and for EPS of US$1.55 and US$1.60. And the critics came out in full force—both in traditional and online media—warning of soaring inventory levels, increasing competition, and slowing growth.

Before I delve into the company's quarterly results, I want to share with you the highlights from lululemon's 2011 video annual review, which the company showed during its annual general meeting last week. I encourage you to watch it, as I think it captures the essence of lululemon's greater purpose, as well as its incredible culture and community. The video, set to the backdrop of beautiful Vancouver (my hometown), is playful, fun, and inspirational. Christine Day, president and CEO, opens up the video talking about how lululemon is "building a culture of personal development and accountability... building leaders around the world"and emphasizes, "It's our beliefs, our values, living the lives we love." She then goes on to share, "It is the dreams of the people who work here, how we treat our guests every day, how we work with our factories, our vendors, our ambassadors, and our communities to achieve these results that truly matter."[26]

As we enter the age of the social renaissance, companies like lululemon that have established a strong level of trust and built up positive social capital with their customers, employees, and business partners will be able to really leverage the higher level of connectedness and create new generative moats. lululemon's unique ambassador program, which the company describes as a "partnership, friendship, and authentic relationship," is the secret to the company's success. In addition to building strong advocacy for lululemon on a grassroots level, the ambassadors provide lululemon with insight into what products to make, what sports are cool, and what is important in each community. This, combined with the time lululemon invests in getting feedback from athletes and guests, creates a very social collaborative environment. Although many companies will try to become more transparent, authentic, and engaging with their customers in their social media efforts, as Frank Eliason writes in his book *@Your Service*, "The creation of a relationship hub does not happen overnight, nor does it start with a social media campaign. It first begins with your own employees and your company culture."[27]

Back to the quarter: if the critics had taken the time to listen to the analyst conference call, they would have discovered the story behind the numbers and realized that the sky was not falling and that, in fact, the outlook for lululemon has never been better. The company's reported gross margin of 55% is right in line with the company's long-term target, and it was actually down only 230 basis points after you adjust for last year's nonrecurring tax benefit. The company benefited from leverage of fixed occupancy and depreciation costs, but this was slightly offset by a bit of inflation and a more normalized level of markdowns. The primary culprit was higher raw material and labor costs resulting from management's strategic decision to invest in product innovation in terms of new fabrics and blending fabrics on garments, which requires more complexity and construction. And yes, the inventory is up 67% from a year ago, but this is positive, as the company's store base is up 30% (it ended the quarter with 180 stores) and it was highly under-inventoried a year ago.

What I find fascinating about lululemon is its conservative yet innovative approach to growth in terms of seeding markets by launching showrooms to test out new store locations, capsules to test out new product lines, and e-commerce to test out new geographic markets. I just finished reading Grant McCracken's incredibly insightful book *Culturematic,* in which he writes, "If you want to innovate as Google, Apple... you should follow a couple of rules: Do not look for big ideas. Seek small ideas that can grow. Fail fast. Fail often. Keep learning and never give up."[28] This describes lululemon's capsule approach to product expansion. For example, last quarter, the company tried out its new swim and commuting lines, and although it has decided to drop them for now, the company believes in this beta approach and is increasing the cadence of new capsule launches from two to three last year to eight this year. As Christine Day explained in the conference call, "These are small, controlled bets, one time, where we take the learnings, always looking for that anchor piece that we can take forward."

It is rare to find a president and CEO who will stand up and declare to analysts, as Day did, that "revenue upside opportunity is limited until Q4, as we have focused our product team on innovation for the future in favor of chasing near-term revenue dollars." But as she went on to explain, "Longer term, making sure that we have the quality, that we have the right amount of scarcity, that we're innovating what's needed as a market leader from a brand and long-term growth story is a position we'd rather be in." It's interesting, as lululemon is very similar to LinkedIn, which cautions in its prospectus, "Our core value of putting our members first may conflict with the short-term interests of our business."

It is ironic that the critics warn that lululemon's growth is slowing—the reality is that the company is on the verge of international expansion. In April, the company opened its first showroom in London, England, which is apparently performing better than the company's US showrooms, and it plans to open its Hong Kong showroom at the end of the third quarter. To help further seed these markets, lululemon will be launching country-specific e-commerce sites for the United Kingdom and Hong Kong later this year, following the May launch of its Australian e-commerce site. The company is starting to reap the benefits from its strategic investment a year ago to bring its e-commerce platform in-house. Its online revenue grew by 179% to US$38 million and now represents 13.5% of total sales, nearly double from 7.4% a year ago. The company's SG&A (selling, general, and administrative) margin actually declined by 160 basis points to 25.6%, as the lower cost structure of its online platform more than offset investments the company made to build its pipeline, infrastructure, and operations. As stated by Kathryn Henry, its chief information officer, in the video, "It's about the global expansion—e-commerce, showrooms, stores... the complexity in the supply chain in order to support bringing products into Europe, Asia, and these new markets we are going to bring lululemon into."

Unlike James Dean, lululemon is a rebel *with* a cause.

***

It's important to remember that, as I mentioned earlier, the rebel-with-a-cause DNA imprint of a company can start to fade if the rebels themselves decide to become less involved or leave. This is what happened with lululemon after its founder, Chip Wilson, decided to retire from his executive post as the company's chief innovation and branding officer in January 2012. A little over a year later, in March 2013, lululemon faced its sheer-pants fiasco (whereby it was forced to recall 17% of its black pants because the luon material used to make them was too sheer), which led to the surprise departure three months later of its president, Christine Day. The company's troubles deepened in November 2013 when Wilson blamed lululemon's customers for the see-through pants issue during a Bloomberg TV interview, stating that "some women's bodies just don't work for [lululemon's yoga] pants." This sparked customer outrage and resulted in huge loss of advocacy, which led to the resignation of Wilson as chairman and the appointment to CEO of Laurent Potdevin, the former president of TOMS and CEO of Burton Snowboards.

In June 2014, Wilson, who was still the majority shareholder of lululemon, raised his concerns that the company was no longer aligned with the core values of product and innovation on which he founded it. Wilson voiced these concerns even more loudly two years later in an open letter to lululemon shareholders dated June 1, 2016: "Unfortunately, lululemon has lost its way and I believe a call to action is needed. I feel strongly that our current Board and management team must clearly articulate and execute a strategy with urgency towards regaining lululemon's competitive advantage and profitable growth and they must take responsibility."

In the letter, he pointed out how lululemon has significantly underperformed compared to its peers since Potdevin took over as CEO in December 2013, with its stock price having declined by 8% compared with gains of 45% and 79% for Nike and Under

Armour in the same period. Wilson went on to state: "We have witnessed a dramatic erosion of the Company's unique culture and capability that empowered and embraced innovation, technology, and product development. This culture and what it accomplished fueled our brand positioning, margins, and growth."[29] Personally, I think lululemon has been busy playing defense in recent years and, as a result, missed out on the opportunity to embrace the consumer wearable-technology space (like Nike and Under Armour have), which would have been a natural fit for the company to advance its movement to "create components for people to live a longer, healthier, and more fun life."

Regardless, I continue to admire the rebel-with-a-cause founders like Chip Wilson, Steve Ells, Howard Schultz, and John Mackey, who have built not just companies but movements. In the nomenclature of Joseph Campbell, these are the tricksters who relish disruption of the status quo and turn the Ordinary World into chaos by refusing to play along with the traditional myopic corporate game of maximizing shareholder value by exploiting their stakeholders.

6

Shift to Social Consciousness Elevates Advocacy

Time to Look beyond Responsible Investing

Working as a sell-side equity analyst was my dream career—I got paid well to do what I love, for people who love what I do, and I had the luxury of working with a team of sales and trading professionals to distribute and market my research. But when I left the corporate world and set out on my own, I soon realized that although I was far more passionate about the more qualitative-based research I was doing because it truly aligned with my values, I was missing a critical component: it was difficult to find people who loved what I was doing. As Pip Coburn, a former global tech strategist for UBS Investment Bank and the founder of Coburn Ventures, observes in *The Change Function*, "Culture is missing from the work of most economists."[30] And based on my experience, I would argue that culture is also missing from the work of most investment professionals.

I saw social media as the catalyst leading to a structural shift in consumer behavior from passive to active as people sought to buy products and services from companies that reflected their own personal values. Over the next decade, I figured this rising base of empowered consumers might start to look beyond their wallets to their piggybanks. This could lead to a structural shift in investing from passive to active as investors began to scrutinize their portfolio holdings and looked to invest their hard-earned savings in companies whose corporate values resonated with their own personal values and had a greater purpose. As stated by Marc Lane in the 2006 article "Advocacy Investing Catnip for Wealthy Clients," "You're talking to clients not just about their money, but about their values—the kind of world they want to pass on to their grandchildren."[31]

But in retrospect, I had essentially decided to cannibalize my existing professional relationships and the reputation I had established for myself as a top-ranked business trust analyst. My attempts to share my "heart and soul" advocacy investment thesis with my former institutional clients (whose mandate was to invest in value-oriented stocks with a high dividend yield) were futile; they had absolutely no interest in looking at high-multiple growth stocks like lululemon, Chipotle Mexican Grill, Whole Foods Market, and Starbucks. And I was actually stabbing them in the back with my social revolution thesis. Having alienated the majority of investors, I figured the socially responsible investing (SRI) community might be receptive to my research. But I should have trusted my instincts: I knew from the minute I started talking with people in the SRI community that they were not my tribe.

I don't usually attend presentations hosted by the CFA Society Vancouver—quite honestly, I find them boring and too academic. However, when I received an invitation to attend a CFA breakfast in early March 2013, cosponsored by the Shareholder Association for Research and Education and discussing the UN Principles for

Responsible Investment (PRI), I eagerly registered in the hopes that it might advance my thinking. When I had begun looking into the concept of stakeholder equity two years earlier, I started to go down the SRI path and did a lot of due diligence in this growing area.

However, since my passion lies in researching companies with heart and soul, I was disappointed to discover that most of the research was focused on how investors can incorporate ESG (environmental, social, governance) factors into their analysis from a fiduciary perspective, to reduce exposure to companies that generate negative externalities. There seemed to be too much emphasis on the E and the G factors in the equation, and the S was focused on human rights and supply chain issues. Given that consumer spending drives nearly three-quarters of GDP growth, and the importance of human and intellectual capital, I was mystified as to where the research was on the C (consumer) and other E (employee).

At the CFA event, Dr. Wolfgang Engshuber, chair of the PRI, spoke about how the PRI was launched in April 2006 as a partnership between the United Nations and investors to promote responsible investing. In the preceding seven years, the PRI had grown its membership base to 1,150 signatories, representing US$32 trillion in assets under management, equal to 15% of the world's investable assets. The list of signatories included 300 pension funds and 600 investment managers; the remainder were service providers. When I spoke with Dr. Engshuber after his talk, he told me how he had retired early after a 25-year career at Munich Reinsurance so he could make a positive difference in the world of investing.

The six principles of the PRI are voluntary and aspirational and they call for signatories to incorporate ESG issues into their investment analysis and decision-making processes, and to proactively seek disclosure from the companies they invest in. As Engshuber pointed out, markets are not efficient, since most investors do not take into consideration the potential impact on a company's future cash flow from nonfinancial components (e.g., ESG factors) and they naively assume that risk follows a normal distribution—which is not always the case. This reminded me of Nassim Nicholas Taleb's latest thinking, which he shares in his book *Antifragile: Things That Gain from Disorder*: that unlike black swan events, which are unpredictable by nature, "you can state with a lot more confidence that an object or a structure is more fragile than another should a certain event happen."[32] As risk management is one of the fiduciary duties of investment managers, it seems that one of the best ways to assess the fragility of the economies, industries, or companies is by looking at ESG risk factors.

Engshuber spoke to how value creation is changing in the 21st century, bringing new sources of financial risk and opportunity because of the increasing importance of environmental issues such as climate change, the increasing complexity of technology and industrial safety, and the rising reputational risk stemming from global supply chain risks. If anything, though, I found his presentation a bit too understated. Maybe I am overly radical in my thinking, but not once did he mention the role social media is playing in making the world more transparent and connected. I strongly believe that as the truth about how companies treat their various stakeholders (whether it be their employees, customers, business partners, communities, or the environment) is exposed, it will create a high level of fragility for companies that derive their competitive advantage by exploiting their stakeholders, since, for the first time ever, the exploited have a voice and are empowered to join together and fight back.

One of the biggest obstacles facing the PRI is the prevailing short-term mindset of investors. Most investors are narrowly focused on companies' next quarterly earnings and value companies based on next year's valuation ratios, so they miss the bigger picture and do not consider how the externalities created by a company will impact its future cash flows and risk/growth profile. The other challenge is the lackluster historical performance achieved by SRI funds (according to a 2012 Deutsche Bank study, 88% of SRI funds showed neutral or mixed results), which has spawned investor indifference and cynicism toward the concept of sustainable or responsible investing.

To generate alpha (i.e., excess return) within the sustainable/responsible investing framework, investors need to move beyond their fiduciary mindset of exclusionary screening and proactively seek companies that are actually creating positive externalities. They also need to expand their focus to the two key stakeholder groups that are largely ignored by the ESG framework: consumers and employees. In the new social era, an age characterized by increasing levels of interconnectedness and individual empowerment, the companies that will outperform are those with the greatest sense of humanity and that display a high level of transparency, authenticity, and engagement in terms of their greater purpose, core values, culture, and community.

Going back to the CFA meeting, I believed heart and soul stocks offered an attractive investment opportunity for institutional investors with an SRI mandate. Instead of just focusing on SRI from a fiduciary or social change perspective and screening out the companies that fail the environmental, social, and governance (ESG) criteria and that generate negative externalities, I thought it made sense to look for companies with a greater purpose using my Four Cs criteria (as outlined in Chapter 4). Heart and soul companies are an ideal fit for socially conscious investors, as they have a greater purpose beyond profit, a set of authentic core values that will serve to morally guide them, and a balanced stakeholder network. They contribute positively to society and create positive externalities, since they are desirable employers, are actively involved in making their communities better places to live, and care for the environment. However, I received a very cool reception to my ideas, which I had posted on my blog and brought up in discussions—most of the SRI community was entrenched in the status quo of negative screening.

Millennials: Emergence of the Socially Conscious Growth Investor

In early 2014, I came across some wise words of Heidi Grant Halvorson and E. Tory Higgins in their book *Focus: Use Different Ways of Seeing the World for Success and Influence*: "Promotion focus is about maximizing gains and avoiding missed opportunities. Prevention focus, on the other hand, is about minimizing losses."[33]This was a moment of awakening for me, as it made me realize that I have a strong promotion-focus mindset, which explains why I thrived in my fast-paced career as a sell-side equity analyst but felt unfulfilled working as a portfolio fund manager. It also explains why I did not feel a kinship with anyone in the social responsible investing community, since everyone I met had a strong prevention-focused mindset, whereas my heart and soul thesis evolved from my promotion-focused DNA.

It was also during this period that I received an invitation on LinkedIn from Rhoden Monrose. I usually ignore LinkedIn invitations from people I don't know, but I was intrigued when I noticed in Monrose's profile that he had spent the previous five years working on Wall Street as a derivatives trader. After a few e-mails back and forth, we had a phone call, and I found it fascinating to learn that this 27-year-old had recently left his lucrative career to found CariClub, to "build a more socially responsible Wall Street, from the ground up." What was even more intriguing was Monrose's observation that "for millennials, social responsibility is their first language… these things are expected, they are a nonnegotiable." I started thinking about this and realized that it was this wave of socially conscious and empowered millennials that would be an influential force on corporate America and Wall Street over the next few decades as they sought to align their values with those who they bought from, worked with, did business with, and invested with. And most importantly, as their baby-boomer parents started to retire and the transfer of wealth began, this new socially conscious generation of some 80 million (between the ages of 21 and 36) would be looking to invest their money in alignment with their values and beliefs.

Up until this point, the only distribution platform I'd had for my articles was my Brady Capital Research website, which had received a total of 400 views in the first two months of 2014. So when LinkedIn announced it was doing a pilot launch of its publishing platform (which prior to then had only been available for Influencers), I eagerly signed up as one of the first 25,000 members to gain early access. I was a bit nervous about publishing my research for everyone to see, but I had been blogging for two years by this point and welcomed the opportunity to gain a wider reach. And so, in March 2014, I published my first article on LinkedIn's publishing platform. I gave it the title "Millennials: The Emergence of the Socially Conscious Growth Investor." It was a thrill to see my post go live, and I was excited when the number of views climbed over 100 in the first hour. It was incredible to think that I was now able to reach professionals all over the world. And, even better, if they "liked" my post, I could look at their profile, thus creating bridging capital.

A study by UBS in early 2014 found millennials to be the most fiscally conservative generation since the Great Depression because they "experienced the financial crisis and all the market volatility and job security issues that came with it—early in their careers."[34]However, risk tolerance is still a function of age, as evidenced by the fact that 29% of millennials surveyed in the UBS study had an aggressive or somewhat aggressive risk tolerance level, versus only 15% of baby boomers. Given millennials' young age, their investment objectives will be different from that of their parents, as they will be seeking growth in addition to safety of principal and income. And given the socially conscious DNA of millennials, those of them who are seeking capital appreciation will be looking to invest in innovative companies that are disrupting the status quo and doing social good. I believe this will lead to the emergence of a new class of investor: the socially conscious growth investor. And the socially responsible investing (SRI) community is ideally positioned to meet the needs of the next generation of investors: the millennials.

"Business as usual is changing. While once companies saw sustainability issues as risks to be managed, many now also see sustainability as a source of innovation that drives growth and profitability." These words appeared in *The Value Driver Model: A Tool for Communicating the Business Value of Sustainability,* a paper released in December 2013 by the UN Global Compact and PRI initiative. It highlights the recent advancement in their thinking about sustainability in terms of growth opportunities versus risk management. Notably, it is a radical departure from the ESG-centric focus of their previous papers, as it discusses the business value of sustainability in terms of driving growth, improving productivity, and reducing company-specific risk factors.

I was happy to see their mindset shifting away from a prevention focus of how to minimize losses by incorporating individual ESG factors into their analysis from a fiduciary perspective in order to reduce exposure to companies that generate negative externalities. It was now moving toward a promotion focus of how to maximize gains by advocating a more holistic mindset through looking at all stakeholders as a source of sustainable value creation by finally including the C (Consumer) and other E (Employee) in the equation. A good analogy is to envision the SRI community advancing up Maslow's Hierarchy of Needs pyramid. Instead of fixating on the bottom two layers (physiological and safety) as a means to reduce their risk exposure to companies with negative social capital, they are now starting to look at the top three layers (belonging, esteem, and self-actualization) as a means to maximize investment growth opportunities in companies with positive social capital.

In addition to focusing on traditional "best in class" companies, investors should look to invest in the companies that are best positioned to capitalize on competitive sustainability as a means to drive growth, improve productivity, and reduce company-specific risks: purpose-driven companies with a stakeholder-centric mindset that are focused on long-term sustainable value creation. As Marc Gunther advises in *Faith and Fortune*, "A willingness to think about the long term sets exemplary companies apart from those that conduct their business with an eye on the day's closing stock price."[35]And as Thor Muller and Lane Becker state in *Get Lucky*, "A real, honest-to-goodness purpose is a core conviction that remains true no matter the external changes to technologies, tastes, or stock prices. It's meaningful because it reflects the unique perspective of the firm and its members."[36](As an aside, I met Lane Becker in the summer of 2008 in San Francisco when I tagged along with Greg, who was meeting with Becker to find out more about his company, Get Satisfaction, which at the time was building a disruptive online platform to empower consumers with product complaint issues.)

Following her stint as CEO of lululemon, Christine Day, in early 2014, took over the helm of Luvo, a start-up aiming to bring healthy food to the frozen-food aisle. Day discussed her thinking behind this decision in a CNBC interview, explaining, "For me, it's all about purpose-driven companies." The challenge for the SRI community will be to figure out these rebels with a cause, as their transformational leaders would rather invest their time and energy on disrupting the status quo to advance their movement than file corporate social responsibility (CSR) reports. While ESG information can be helpful to risk analysis, it is not sufficient for identifying companies that practice capitalism in a manner that benefits all stakeholders, including investors. Instead of waiting to invest in these companies until they file a formal CSR report, investors may wish to assess companies' levels of transparency, authenticity, and engagement by using their corporate Facebook, Twitter, and LinkedIn social media sites as real-time listening posts. And crowdsourced-based review sites such as Yelp, TripAdvisor, and Glassdoor can provide investors with additional insight into customer and employee perceptions of these companies.

The new social era of transparency, connectedness, and stakeholder empowerment is creating new social value creation opportunities of advocacy, learning, and collaboration/co-creation. The companies best positioned to capitalize on these opportunities to build long-term, sustainable value are those with a greater purpose, authentic core values, and a strong corporate culture and that are focused on generating positive social capital (i.e., shared value or positive externalities) for their customers, employees, suppliers, communities, and the environment. Although positive social capital is an intangible asset that, unlike goodwill, does not show up on a balance sheet, investors need to factor it into their analysis and valuation process, since it will lower companies' risk profiles and enhance their growth profiles. As Diana Rivenburgh states in *The New Corporate Facts of Life*, "Forward-thinking businesspeople are replacing the old shareholder-centric mindset with a much more inclusive shared-value one." Companies and investment firms looking to attract millennials would be well advised to take heed of this maxim: "You are what you do, not what you say."[37]

7

The Value of Connection

Social Networking Sites Democratize Data

I had an e-mail exchange in 2007 with one of my institutional clients, who was puzzled and somewhat suspicious of the mysterious invitation he had just received from me. "What is LinkedIn?" he inquired. I thought about it for a second and responded, "LinkedIn is Facebook for professionals." I had just joined and was excited about the prospect of being able to create a virtual Rolodex of all my professional contacts. However, at that time, LinkedIn had only 15 million members globally and, judging by the lack of enthusiasm I received, the finance industry was apparently not an early adopter.

As LinkedIn gained members and respect among professionals over the next four years, it became an invaluable tool for me, allowing me to reconnect and stay in touch on a professional level with clients, former colleagues, classmates, and friends. But I didn't begin to appreciate its true value until the fall of 2011, around the time I was preparing to launch Brady Capital Research, when I started to think about how I could use LinkedIn to reach out to professionals around the world who might be interested in my research.

At this point, I had pretty much finished writing *Social Capital: A New Strategic Play for Investors*, and I was eager for a new challenge. Intrigued by LinkedIn's burgeoning popularity and sensing that there were things I could learn, I decided to take a closer look at the company, which had gone public the previous spring. I began by listening in to the 3Q11 conference call and soon became fascinated by the company's business model and its value creation potential. After a month of doing due diligence on LinkedIn, I created an outline for a report on LinkedIn and started to go through my business strategy books to gather insights into the company. Something told me that this might be my next lululemon stock pick, so I added LinkedIn to my portfolio, buying the stock at US$68.66.

I spent countless hours trying to figure out LinkedIn's cost structure and the company's revenue drivers so that I could model its future free cash flows and come up with a realistic discounted cash flow value. However, as the company was already well covered by the Street (by 21 analysts back then, according to FactSet), I didn't think I could add any incremental value on the accounting or numbers side of the equation. But where I did think I could add value was on the qualitative strategic side.

In the ensuing nine months, I focused my research on how LinkedIn capitalized on the emerging democratization of data to empower professionals to take active control of their careers by enabling them to make connections to build bonding, bridging, and linking capital. Although linking capital was the transaction and value-capture part of the equation that investors care about, I was excited about the potential for LinkedIn's bridging capital power to spawn a second renaissance in intellectual capital. What I learned provided me with the insight to unlock the secret value social driver leading to abundance of supply: connection. Allow me to explain how I came to that point.

In January 2012, I connected with Rod Beckstrom, who cowrote the visionary book *The Starfish and the Spider,* published in 2006. I shared with Beckstrom how the insights in his book had given me the conviction in January 2008 to go against the Street and downgrade Yellow Pages to a sell. (Incidentally, this intrigued him and he invited me to share my story, which appeared in the April 2012 issue of *The Beckstrom Starfish Report*). Beckstrom foreshadowed how LinkedIn, with its starfish-like DNA, possesses the power to disrupt the business relationship management market when he wrote: "Spider organizations weave their webs over long periods of time, slowly amassing resources and becoming more centralized. But the starfish can take over an entire industry in the blink of an eye."[38]

About a month after I'd connected with Beckstrom, I read the short but powerful book *Gutenberg the Geek,* by one of my favorite business strategy authors, Jeff Jarvis, who provided me with insight into how the democratization of data was leading to the new social value driver of connection. Jarvis posits: "The internet... is not just a means of data exchange but of cultural exchange. It is not, in my view, a medium but instead a connection machine."[39]

It's interesting how you can spend so much time working on something and it's not until the last second that you get a revelation that brings everything together. This is what happened to me. I'd just finished the final draft of my research report on LinkedIn and was sharing with Greg how LinkedIn's new Talent Pipeline platform allowed companies to keep their applicant data current. I did not think much of it until he explained to me that this was one of the biggest challenges facing companies, and that he actually struggled with this same problem back in the mid-1990s when he created Vision2Hire, one of the first web-based applicant tracking systems. And the next morning it dawned on me: through the "power of we," LinkedIn had discovered the holy grail of business relationship management—how to transform candidate pools, business prospect leads, and business customer bases from depreciating into appreciating assets.

On May 16, 2012, I published my research report *LinkedIn: Disrupting by the "Power of We."* Since LinkedIn went public on May 19, 2011, at US$45 per share, the value of the stock had increased by 146% to US$110. With a then current valuation of over US$11 billion, the stock was trading at the astronomical multiple of over 150 times earnings. LinkedIn was coming under attack from short sellers, who decried the stock's excessive valuation, denounced the company as being little more than a glorified online Rolodex, and questioned the company's ability to monetize its social networking platform. However, I believed there was much more to the LinkedIn story than the numbers revealed.

The Three Forms of Social Capital

LinkedIn is a massive global network of people, with each professional member consisting of a single node. More important than the nodes are the ties established between members, and the strength of these ties. Mark Granovetter, in his seminal 1973 paper *The Strength of Weak Ties,* proposes that professional career opportunities are more strongly influenced by casual acquaintances than by those who you know well, as people with whom you have weak ties tend to move in different professional circles.[40] And as Robert D. Putnam states in his 2000 book *Bowling*

Alone, "Social capital refers to connections among individuals—social networks and the norms of reciprocity and trustworthiness that arise from them."[41] This principle underlines LinkedIn's "power of we": it enables professionals to form weak ties through bridging capital. This is discussed in Guy Champniss's book *Brand Valued,* which explores the concepts of linking capital (the formation of links between vertically- or horizontally-separated groups), bonding capital (the formation of strong ties), and bridging capital (the formation of weak ties).

LinkedIn's vision to "create economic opportunity for every professional in the world" is about building human capital for its members by helping them link to potential employers, clients, and business partners. Linking capital is the transaction and value-capture side of the equation that forms the basis of LinkedIn's three business lines: Talent Solutions (formerly called Hiring Solutions), Marketing Solutions, and Premium Solutions. LinkedIn builds bonding capital by enabling members to connect with former colleagues and others they know on a professional basis. These are strong ties, as there is a high degree of homogeneity between members' first degree connections, and they tend to reinforce shared practices, norms, and similarities. However, LinkedIn is much more than just a virtual Rolodex. Through innovative products such as LinkedIn Today, Updates, and Groups, LinkedIn facilitates opportunities for interactions that create and transfer knowledge between members. As John Hagel and his coauthors observe in *The Power of Pull,* we can no longer rely on "stocks" of knowledge, since they are diminishing in value more rapidly than before, so we need to participate in relevant "flows" of knowledge.[42] LinkedIn allows its members to participate in these flows of knowledge by leveraging their networks of trusted professional contacts.

LinkedIn builds bridging capital for its members, enabling them to form weak ties with professionals two or three degrees of separation away and to join groups. LinkedIn's bridging capital power comes from its ability to facilitate what coauthors Stephen M.R. Covey and Greg Link refer to as "smart trust," by providing transparency into its members' credibility in terms of their competence (e.g., profiles provide insight into the level of competence by listing capabilities and track record of results) and character (e.g., LinkedIn's platform enables its members to check the integrity and intent of members two or three degrees of separation away through shared contacts).[43]

LinkedIn is creating what visionary thinker Frans Johansson calls the "Medici Effect" and is spawning a second renaissance in intellectual capital.[44]By facilitating trust and allowing its members to build bridging capital and share their expertise through updates and published posts, LinkedIn breaks down the barriers between professional disciplines, age groups, and geographies, allowing its members to find one another, learn from one another, and form heterogeneous groups. These creative collaborative efforts are resulting in the combination of concepts between multiple fields, generating intersectional ideas, and increasing the quality and frequency of innovation. This is leading to a second renaissance as new industries, markets, and ways of making a living are formed.

LinkedIn's power to create both bonding and bridging capital for its members provides it with a distinct competitive advantage over leading social networking platforms such as Facebook and Twitter. Facebook's disadvantage is that it does not have strong bridging capital across networks—the nature of its social network is based on bonding capital with families and friends. Twitter's network is based on bridging capital, with members allowed to "follow" anyone they want. However, Twitter's platform does not facilitate trust because it does not provide the means to assess the competence or the character of its members. LinkedIn's ability to create bridging and bonding capital on a global basis provides it with a strong competitive advantage over smaller localized professional networking companies such as Xing in Germany and Viadeo in France.

It was with a sense of relief yet sadness that I heard the news on June 13, 2016, that Microsoft was acquiring LinkedIn for US$196 per share, equating to a purchase price of US$26.2 billion. The relief came from reaching the end of the wild rollercoaster share-price ride I had been on the past four years: after reaching a peak of US$269 per share in February 2015, LinkedIn's share price had crashed to nearly US$100 in February 2016, after missing analyst expectations. But I felt sadness also: although LinkedIn claims its mission to "connect the world's professionals to make them more productive and successful" is aligned with Microsoft's mission to "empower every individual and organization in the world to achieve more," I'm concerned LinkedIn will lose its rebel-with-a-cause attitude. Although I understand LinkedIn's desire to seek refuge from the myopic clenches of Wall Street, I wish it had been able to go private on its own. I'm worried that LinkedIn may become more focused on identifying ways for Microsoft to leverage and monetize its structural asset base than on spawning a second renaissance in intellectual capital.

The Social Renaissance

CEOs are looking beyond the benefits of connected supply chains and more integrated back-office systems. Their focus is shifting to the power and potential of recent advances in social media and analytics to reimagine connections among people—whether that's customers, employees, partners, investors, or the world at large.

I read these opening remarks by Ginni Rometty, president and CEO of IBM, in the 2012 IBM CEO study and pulled out my highlighter, excited to dive into the report.[45] As the title suggests, the study, based on face-to-face conversations with over 1,700 chief executive officers in 64 countries, is all about how social media is bringing in a higher level of connectedness.

It was now June 2012, and for about two years I had been mulling over the concept of how social media would act as a catalyst to structurally shift companies' risk and growth profiles. Interestingly, in December 2011, *Time* magazine named "The Protester" as its Person of the Year. And in the new year of 2012, the Stop Online Piracy Act (SOPA) protests demonstrated how social media allows grassroots movements to build momentum at a level of intensity and pace never witnessed before, by allowing individuals to self-organize and by increasing their clout and bargaining power. But I did not think investors had yet recognized how the social revolution would be negative for companies that derived their competitive advantage through exploiting their customers, employees, business partners, communities, and the environment.

The reality is that over the years, these exploitative companies have built up distrust and negative social capital with their stakeholders. Unlike goodwill, negative social capital (i.e., badwill) is an intangible that does not show up on the balance sheet, so investors have largely been able to ignore this inherent liability. However, social media empowers these exploited stakeholders to fight back. The crux of the matter is that the social revolution will lead to an erosion in the competitive moats of these exploitative companies, resulting in an increase in their risk profile and decline in future growth opportunities.

If you recall, I launched Brady Capital Research to focus on companies working to make a positive difference in the world. So I got to thinking about the positive side of social capital and the IBM connectedness study. Hmm, maybe social media is ushering in the social era—and in addition to giving rise to the social revolution, it is also giving rise to the dawn of the social renaissance.

It is times like these that I wish I'd studied humanities at university instead of business. I am a huge fan of the Renaissance. I have visited Florence many times, I like to paint and draw, and one of my favorite books is Michael J. Gelb's *How to Think Like Leonardo da Vinci*. What I found interesting when I started reading up on the Renaissance is that the Humanism movement focused on teaching citizens subjects such as grammar, rhetoric, history, poetry, and moral philosophy to enable them to engage in the civic life of communities and to persuade them to act virtuously and prudently. The Humanism movement was a reaction against the utilitarian approach of the Scholastic movement, which focused on preparing men to be doctors and lawyers, with attention to rules and details.

I was thinking about this as it pertains to capitalism today—maybe if we encouraged young people to study humanities instead of business and to pursue their real passions instead of pursuing just the dollar, we would have a more humanistic form of capitalism. It will be interesting to watch events unfold over the next decade. As empowered customers, employees, and business partners start to revolt against companies that have been able to get away with exploiting them, I have no doubt that we will see their competitive moats erode. This will lead to a natural attrition of the bad companies, raising the morality level of capitalism. But what I think will lead to the true enlightenment of capitalism is the social renaissance.

In the coming era of the social renaissance, companies that have established a strong level of trust and built up positive social capital with their customers, employees, and business partners will be able to really leverage the higher level of connectedness and create new generative moats. These companies will then be able to capitalize on their existing authentic and strong relationships by leveraging the high level of enthusiasm and support for their brand and greater purpose. This will enable them to further advance their movement by strengthening existing relationships and attracting new supporters. In addition, the increase in the flow of knowledge could lead to new collaborative partnership efforts, increasing the quality and frequency of innovation.

The concept of a higher level of connectedness sounds promising, as do IBM's three key recommendations that companies empower employees through values, engage customers as individuals, and amplify innovation through partnerships. However, I suspect very few companies have a high enough level of trust and goodwill in their existing relationships with their customers, employees, and business partners to be able to successfully implement these recommendations. In fact, this higher level of connectedness could accelerate the erosion of the competitive moats of exploitative companies and lead to their hastened downfall. The reality is that the companies poised to really prosper during the social renaissance are the ones that have a greater purpose, and that are transparent, authentic, and engaging in terms of their core values, culture, and community. The bottom line is that the social era is fostering the rise of two conspiring forces, leading to the enlightenment of capitalism: the social revolution and the social renaissance.

The Seed, Soil, and Roots of a Corporate Ecosystem

In early 2013, I read *The Story of Purpose,* in which author Joey Reiman declares: "Beyond wealth creation and shareholder value is an unlimited resource to create a better world and a lasting legacy for businesspeople around the globe. It's called purpose."[46] Inspired by his words, I tweeted out: "@JoeyReiman: Just read *The Story of Purpose* and *Thinking for a Living* this weekend—loved them both—thank you!—Barbara." To my surprise, Joey tweeted me back, and this led to an e-mail exchange and then a series of conference calls with his colleagues at his purpose-driven consulting firm, BrightHouse (which he sold to Boston Consulting Group in May 2015). This inspired the idea for the paper *Why Companies with a Greater Purpose Will Thrive in the New Social Era,* which Joey and I cowrote and published on April 29, 2013. Looking back, I realize that this paper planted the seeds for the third social value driver: collaboration.

Unlike in previous eras, consumers now have access to large amounts of knowledge and a great ease of connecting to like-minded partners to spread their individual message on a more massive scale. Social media exchanges act as catalysts to accelerate the formation of weak and strong ties, leading to the formation of bonding, bridging, and linking capital between and among a company and its various stakeholder constituencies. Social media is turning the silent majority into a revolutionary force of highly connected consumers, employees, and suppliers, and empowering them to influence, expose, and disseminate their views (both positive and negative) on a company's products and services, how a company treats them or other stakeholders, and how a company's business activities impact society and the environment. In addition, these stakeholders can now self-organize and join forces on a common cause to gain support from others and to lobby for social and regulatory reform and change, as well as create new communities to enable learning, collaboration, and co-creation.

This is leading to the creation of the new equity form of social capital. And as the number of individual and corporate users of Facebook, Twitter, and LinkedIn increases; new social exchange platforms like Google+ and Pinterest emerge; and the density of connections within and between the different social networks increases, social capital will, I expect, continue to appreciate in value.

As the shift from material want to "meaning want" progresses, people will look for companies that offer more than just a utility (i.e., functional or financial) or an emotional value proposition. They will be looking for companies that stand for something—companies with a greater purpose. I visualize the stakeholder ecosystem of a company like the root system of a tree, composed of the seed, the soil, and the roots. As Chip Conley advises in *Emotional Equations,* "Knowing where you come from (your roots) will help you know who you are (authenticity) and where you're going (your calling)."[47] And as Joey Reiman wisely observes, "The fruits are in the roots":[48]

> *When you rediscover your organization's true identity—what it stood for, then you will gain fresh insight into the reasons for its existence, its essence, its why—its very soul. It doesn't*

change over time. Yet, too many lose sight of it and do not know how to get it back. Too many companies, in America, become lost souls. So many brands stand for nothing. They have lost their purpose and meaning. To discover what will make your company or brand truly great in the years to come is to discover its history—its WHY—and rebuild from there.

On April 7, 2014, I published the in-depth research report *Zillow: Disrupting the $75 Billion Realtor-Centric Machine*. Since Zillow had gone public on July 20, 2011, at US$20 per share, the value of the stock had more than quadrupled in value to US$91 per share. On the surface, Zillow looked highly overvalued as it had generated under US$200 million in sales the previous year and was trading at an enterprise value (EV) of US$3.4 billion, implying an EV/sales multiple of 17 times.

However, when I dug below the surface to the roots, I uncovered a company whose greater purpose to "lead a revolution in online real estate to empower consumers" had created a thriving ecosystem around its unique business model, with the potential for significant long-term sustainable value creation as it disrupted the US$75 billion traditional realtor-centric machine. For in less than a decade, Zillow had created a thriving ecosystem with significant scale and influence by generating positive social capital (i.e., shared value and positive externalities) for its then community of 800-plus employees, over 70 million unique visitors, some 650,000 real estate agents, four leading media distribution partners, and the US government. To gain insight into how Zillow had been able to achieve this level of scale and influence in less than a decade, investors would need to look below the surface to the seed, soil, and roots of its ecosystem.

The *seed* represents Zillow's greater purpose and DNA. Back in

2006, Richard Barton and Lloyd Frink, inspired by the motive to disrupt the real estate market like they had a decade earlier with the launch of Expedia, planted the seed for Zillow's accessibility-driven movement. According to the company, "The Zillow name evolved from the desire to make zillions of data points for homes accessible to everyone. And, because a home is about more than just data—it is where you lay your head to rest at night, like a pillow—'Zillow' was born." The founders must have been inspired by Dr. Seuss's *There's a Wocket in My Pocket,* in which the last line reads: "But the ZILLOW on my PILLOW always helps me fall asleep."[49]They saw the potential to launch an Internet-driven revolution to bring transparency to the realtor-centric US real estate market, which still operated from the antiquated MLS platform.

The *soil* represents Zillow's core values and culture. This is where the seed of the company is cultivated. Unlike most companies, which are shareholder-focused and care only about short-term results, Zillow is committed to its key stakeholder (the consumer) and is focused on long-term value creation. Ironically, Zillow actually states this as a risk factor in its annual report:[50]

> *We will in the future forgo certain expansion or short-term revenue opportunities that we do not believe are in the best interests of consumers, even if such decisions negatively impact our short-term results of operations. In addition, our philosophy of putting consumers first may negatively impact our relationships with our existing or prospective advertisers. This could result in a loss of advertisers, which could harm our revenue and results of operations.*

Zillow is a data- and technology-driven company that advocates a culture of winning, teamwork, collaboration, creativity, and innovation focused on its core values of transparency and power to the consumer. As Spencer Rascoff, CEO of Zillow Group, shared in his October 19, 2015, LinkedIn article "How to Create a Culture of Transparency," "Everyone at Zillow Group is aligned behind one mission: 'Power to the People.' There's a lot of academic research about why it's advantageous to have a mission-driven culture, particularly with millennials who are trying to connect with something beyond a job; they want a purpose, not just a paycheck."As evidence of the strength of Zillow's corporate culture, the company has a 4.3 out of 5.0 rating on Glassdoor (based on 430 reviews as of August 2016), with a 96% approval for Spencer Rascoff.

The *roots* represent Zillow's community of employees, consumers, real estate professionals, and business partners, which reside in the soil and sprout from its seed to form its stakeholder ecosystem. A good measure of the reach of a company's stakeholder ecosystem is its ratio of followers to employees on LinkedIn, as professionals usually follow a company if they are interested in working for or with it in some capacity. For example, as of August 2016, this ratio for Zillow Group is 25:1 (over 60,000 followers to just over 2,400 employees). Interestingly, Rascoff has an even broader personal social reach, with over 220,000 followers on LinkedIn.

I was in a state of euphoria in the months after the April 7 publication of my research report on Zillow, as the company's stock price proceeded to soar by over 75%, reaching a peak of US$160 in July. But this was not to last for long. For the seed, soil, and roots of Zillow's corporate ecosystem were disrupted at the end of July when the company announced it would be acquiring its largest competitor, Trulia, for US$3.5 billion. The competitive landscape shifted even further when News Corp. acquired the third-party player, Move, at the end of September for US$950 million.

But the bottom line is that the companies that will thrive in the new social era of stakeholder connectedness and empowerment are those with a greater purpose, having evolved from being soulless machine-like entities to being more human-like and transparent, authentic, and engaging in terms of their core values, culture, and community. As Don Tapscott and David Ticoll declare in *The Naked Corporation*, "Today's economy depends on knowledge, human intelligence, agility, and relationships inside and outside the firm. The fuel is information, and the lubricant is trust."[51] These companies create positive social capital by displaying a high level of trust, optimizing for long-term relationship building, cooperating with stakeholders, and focusing on reciprocity and corporate social opportunity. And as Umair Haque puts forth in *Betterness: Economics for Humans*:[52]

> *Maybe there are better kinds of companies, which can return more than just profit through better approaches to production and consumption, that can yield more meaningful, durable benefits by trading and exchanging hardier, more enduring, more fruitful kinds of capital... They're demanding organizations that do not just make money, but can begin seeding, nurturing, and harvesting higher order wealth in them, with them, and for them. It is up to those organizations that power, advantage, trust, and returns are already inexorably flowing.*

The stakeholder root ecosystem visually portrays how a company's consumers, employees, and partners use social media to intermingle and intertwine and form relationships among themselves. This results in the growth of new secondary root systems as new relationship domains of consumers-to-employees, employees-to-partners, and consumers-to-partners emerge. This ultimately leads to the creation or destruction of a firm's social capital.

In the new social era, companies are no longer able to derive their competitive advantage by exploiting their customers, employees, and partners or by pursuing activities that generate negative societal and environmental externalities. Traditional economic moats such as being a low-cost producer, possessing intangible assets, creating high switching costs, and benefiting from closed network effects are eroding as stakeholders use social media to criticize, protest against, and defect from companies that profit at their expense. As John Gerzema and Michael D'Antonio observe in *The Athena Doctrine*, "In a highly interconnected and interdependent economy, masculine traits like aggression and control are considered less effective than the feminine values of collaboration and sharing credit."[53]

More importantly, enlightened companies with a greater purpose are finding themselves in an enviable position: they are able to start to build bonding, bridging, and linking capital with their stakeholders. This enables them to form communities that allow them to access the new social value drivers of advocacy, connection, and collaboration. In terms of advocacy, a company with strong and trustworthy relationships based on a shared sense of purpose is ideally positioned to benefit from the positive network effect resulting from word of mouth and interactions among its consumers, employees, and partners.

A company can also use social media platforms to dynamically and interactively reach out to its customers, employees, and partners, and to build highly connected communities focused on the passion for its greater purpose and the products and services it offers. This creates a unique experience and further deepens the level of psychological attachment its stakeholders have for the company. And a company with a greater purpose is ideally positioned to encourage consumers, employees, and suppliers to work together toward a shared goal or vision. Social platforms create a new open forum for participation, facilitating different forms of collaboration and co-creation, ranging from user-generated content to idea generation to open-source to crowdsourcing to mass customization.

As Tom Asacker observes in his book *Opportunity Screams*, "The hard work of value co-creation—viscerally understanding the relationships between the sticks and strategically connecting them to each other—is essential to unlocking the Doors to opportunity and achieving sustained success."[54]

8

The Value of Collaboration

Social Capital: The Secret behind Airbnb and Uber

> *The true promise of a connected society is helping one another.*

—Biz Stone, cofounder of Twitter, in *Things a Little Bird Told Me*

Ironically, although I had been on an intellectual journey for the previous four years, it was only when I undertook a physical journey to New York City (I had traveled east to speak at the CFA Society Toronto event Social Media's Impact to the Investment Process) and used Airbnb and Uber for the first time that the dots finally connected. This inspired me to write the article "Social Capital: The Secret behind Airbnb and Uber," which I published on LinkedIn in June 2014. It ended up going viral, and as of August 2016 it has been viewed by over 396,000 professionals. The article serves as a good summary of what I've discussed thus far and shows how my thinking evolved to what I go on to discuss in the rest of the book.

Social Capital: The Secret behind Airbnb and Uber

When I was 30 years old and single and living in New York City, I had a dream: to return one day with my husband and push our baby in a stroller through Central Park. In May 2014, my dream came true, albeit more than a decade later. But we did not stay in a hotel and travel by yellow taxicab; instead we used Airbnb to book a guy's condo on the Upper East Side and we traveled by Uber to the airport.

How is it that Airbnb and Uber have been able to build thriving ecosystems in just over five years, with such significant scale and influence that they were then valued at US$10 billion and US$12 billion? And how have these companies become such a disruptive force that they are the target of deafening protests from the highly ensconced hotel and taxi industries in cities around the world? Two words: *social capital.*

In this new social era of transparency, connectedness, and stakeholder empowerment, social media exchanges are acting as catalysts to accelerate the formation of bonding, bridging, and linking capital among its stakeholders (i.e., employees, customers, and business partners).

There are three levels of companies that operate in the social economy. And the higher a company moves up what I term the Social Economy Pyramid, the faster the rate of value acceleration, as they are able to achieve a higher level of disruption and access multiple social value drivers. The first social value driver is advocacy, which creates abundance of demand through the democratization of influence. The second social value driver is connection, which creates abundance of supply through the democratization of data. And the third social value driver is collaboration, which creates abundance of both supply and demand through the democratization of physical and human capital.

Starbucks, Whole Foods Market, Chipotle Mexican Grill, and lululemon create abundance of demand through advocacy of a social mission. By creating positive social capital (shared values and positive externalities) for stakeholders, these companies are able to attract people looking to align their values with the companies they buy from, work for, and work with. This leads to the creation of a thriving stakeholder ecosystem. Because social mission companies are founded on movements and follow a blue ocean strategy, they are on the low-end of the disruption scale, as they created new uncontested marketplaces. And their ability to grow revenue and free cash flow is limited by time and capital constraints on the supply side.

LinkedIn creates abundance of supply through connection by building a unique structural asset base that empowers people to connect with one another and build bridging and linking capital through their platforms. By revolutionizing the way professionals manage their reputation and contacts, LinkedIn has been able to create a unique and invaluable structural asset base from the profiles and highly interconnected networks of its now 450 million members. This provides LinkedIn with a low-cost supply of constantly updated data (i.e., raw material) from which to extract value. Although LinkedIn may appear unthreatening, with nascent business lines, it has been able to establish a long tail and is beginning to gain footholds in the low-end markets. The revolutionary power of its consumer-centric marketplace platforms makes it, as Clayton Christensen and Michael E. Raynor deem in *The Innovator's Solution*, a "ubiquitous disruptive force" to be reckoned with.[55] However, LinkedIn is challenged on the demand side: I estimate that its paid customer penetration is less than 2%.

Airbnb and Uber create abundance of demand and supply through collaboration by creating social platforms that facilitate trust and enable individuals to form weak ties (bridging capital) with one another, leading to the personal sharing of assets, goods, and services (linking capital). By accessing an untapped market of nonsuppliers of latent and underutilized personal assets, goods, and services, these companies are not only creating blue oceans of demand but discovering a new frontier below the corporate ocean with no capital or time constraints: the social ecosystem reef. And by creating a long tail in rival assets, goods, and services, these companies are able to directly match supply and demand, and collect a cut of each transaction.

When I stayed at my first Airbnb in May 2014, the company had already processed a total of 11 million reservations and created a compelling accommodation alternative for its guests by offering them a fun way to discover and book unique accommodations with its then base of over 350,000 hosts offering 600,000 home listings. And at that time, Uber operated in over 70 cities in 36 countries around the world. As of August 2016, Airbnb has more than tripled its inventory of home listings to over 2 million, and Uber has more than quintupled its operational base to over 400 cities. These rapidly expanding social-sharing companies are directly attacking the incumbent hotel and taxi industries on both the supply and demand side and are threatening to erode their traditional economic moats in terms of low-cost production, high switching costs, intangible assets, and the network effect.

In terms of being a low-cost producer, the incumbents' high fixed-cost structure (which provides them with process and scale cost advantages) is becoming a competitive disadvantage, as many travelers have grown tired of the impersonal experience of staying in cookie-cutter hotels and of the inefficient and impersonal experience of traveling in taxis. In comparison, companies such as Airbnb and Uber are the ultimate low-cost producers, as they have a close-to-zero marginal cost model, since they have the potential to create infinite supply by empowering individuals to generate income from underutilized personal assets (i.e., property, plant, and equipment such as a house or car, and human capital such as property management or chauffeuring services). In terms of high switching costs, customers are no longer held captive to the incumbents as Airbnb and Uber now present more attractive and personal alternatives.

In terms of intangible assets, the value of a brand name is depreciating as Airbnb and Uber create long tails in travel by replacing artificial institutional trust with social capital. They achieve this by democratizing the tools of production and distribution, and by connecting supply and demand by capitalizing on the filtering efficiency of social network reviews and facilitating trust through dual accountability systems (i.e., both the hosts and the guests rate each other). And in terms of the network effect, unlike hotel and taxi companies that seek to constrain supply to keep prices high, Airbnb and Uber are creating structural assets that appreciate in value as they attract more and more new hosts and drivers (i.e., supply) and travelers (i.e., demand) to their platforms, leading to the ultimate network effect.

I have no doubt that the social economy will transform how we travel, live, work, play, and consume. And the secret to understanding this accelerating tectonic shift starts with social capital.

***

The timing of this article was perfect, as I had recently brought aboard a new summer intern, Nikhil Anand, an engineering professional from India who had just completed the first year of his MBA at the University of Toronto's Rotman School of Management. Together we started to explore the uncharted territory of the peak of the Social Economy Pyramid: the emerging sharing/on-demand economy.

Could LinkedIn Become the Uber for Professionals?

In mid-July 2014, I had a great phone conversation with Tristan Pollock, cofounder of Storefront, about how his company was revolutionizing the commercial retail marketplace and creating a "pop-up retail market" by unbundling retail space in terms of time (unbundling the 5- to 20-year lease term into a daily, weekly, or monthly rental license) and space (partitioning off square footage). This inspired me to publish the article "Could LinkedIn Become the Uber for Professionals?," a collaborative effort with Heather McGowan, an academic entrepreneur and innovation strategist. As a testament to the bridging capital power of LinkedIn, Heather had reached out to me after reading my "Social Capital" article (which you just read above) and, in turn, she shared with me her four-series set of LinkedIn articles, "Jobs Are Over: The Future Is Income Generation."

The basic thinking that inspired "Could LinkedIn Become the Uber for Professionals?" was this: Uber empowers taxi drivers with the freedom to work when and where they want; wouldn't it be great if LinkedIn could enable professionals to do the same? Perhaps LinkedIn could revolutionize the professional talent marketplace and create an on-demand "pop-up professional talent market" by unbundling professional talent in terms of time (unbundle the traditional long-term employment agreement into hourly, daily, weekly, or monthly project assignments) and space (professionals could work virtually for a series of companies instead of for just one company at its office). And just as Storefront matched supply (i.e., retailers, brokers, and landlords that have available space) with demand (i.e., artists, designers, and brands that are looking for space), could not LinkedIn also match supply (i.e., professionals with leveraged skills and available time) with demand (i.e., companies seeking intellectual expertise on a specific project)?

The magic of sharing economy companies like Airbnb and Uber is that by harnessing technology to create a two-sided marketplace, they are able to empower individuals and businesses to monetize their idle or underutilized assets. However, LinkedIn would differ from Storefront and Uber as a sharing economy company because it would focus on the monetization of human capital. According to Thomas A. Stewart in his book *Intellectual Capital,* there are three skill sets when it comes to human capital: (1) commodity skills, (2) proprietary skills, and (3) leveraged skills.[56]

What's interesting is that we are seeing the emergence of a wide range of sharing economy companies that enable individuals with the first two skill sets to generate income by monetizing their human capital. In terms of commodity skills, people are able to earn active income by doing personal tasks (TaskRabbit), providing local tour guide services (Vayable), dogsitting (DogVacay), providing household services (Handy), caregiving (Care.com, UrbanSitter), and doing mobile corporate gigs (Gigwalk). In terms of proprietary skills, people are able to earn active income through cooking (Munchery), tutoring (WyzAnt), teaching (Skillshare), creating (Gumroad, Etsy, CustomMade), and skilled freelancing (Upwork).

These sharing economy companies are accelerating the shift from reliance on the corporation (i.e., job income) to reliance on self (i.e., portfolio of income generation). As Heather McGowan proposes in "Jobs Are Over: The Future Is Income Generation":[57]

> *From 1950–2000, the "job," a discrete event that consumed 35–50 hours per week in a set location with predictable tasks and clear career pathway for forty-some years, provided income generation from the paycheck and the pension... Many people will not have annual salaries or set jobs in the traditional sense, but rather they'll generate income from leveraging and monetizing a combination of their physical assets and talents in an income-generation portfolio.*

But where does that leave professionals like you and me who have a leveraged skill set? Although it might be fun for us to earn extra income by acting as a local tour guide or by dogsitting, I don't think that would be the best return on our assets. As J.C. Larreche discusses in *The Momentum Effect,* although we may be great at originating value through our intellectual capital, we need to find a way to extract that value (i.e., turn our talents into actions), and most importantly, capture that value (i.e., turn our actions into money).[58] And this is where I think LinkedIn could come in.

With its compelling social mission and deep foundation of structural capital and customer capital, LinkedIn is ideally positioned to ascend to the peak of the Social Economy Pyramid and transform the way we work and live. During an interview at the TechCrunch Disrupt conference in San Francisco on September 9, 2013, Jeff Weiner, CEO of LinkedIn, discussed his aspiration to develop the world's first economic graph to digitally represent every professional, company, and higher education institute. His visionary thinking is evidenced by his statement, "Our goal would then be to get out of the way and allow each of those nodes to connect where it can create the most value—or capital, all forms of capital: intelligent capital, working capital, human capital... And in doing so we hope to play a role in transforming the global economy." And when this happens, investors will realize there is much more to the LinkedIn story than the numbers currently reveal.

On October 19, 2015, LinkedIn launched LinkedIn ProFinder, its professional services platform, and began piloting it in San Francisco with three categories (accounting, graphic design, and writing and editing). And mentioned in Chapter 2, it has since expanded its platform to feature US-based experts across 10 professional services categories. As a Canadian, I am hopeful that LinkedIn will soon roll out its platform to professionals located north of the border.

Value Creation: Democratization of Physical and Human Capital

Abundance economy companies enable individuals and businesses to monetize their underutilized or latent physical and human capital by providing a platform that provides what Thomas A. Stewart refers to in *Intellectual Capital* as structural capital and customer capital.[59] Mashing together this concept with the three forms of value creation (value origination, value extraction, value capture) discussed by J.C. Larreche in *The Momentum Effect*, I developed a value creation framework to explain how abundance economy companies create value:[60]

- **Originating** value through physical and human capital.
- **Extracting** value through structural capital.

- **Capturing** value through customer capital

Originating Value through Physical and Human Capital

Value origination is the true source of momentum for abundance economy companies. Consequently, to gain insight into their underlying economic characteristics, we need to look at their source of value origination (rather than classifying these companies by sector). And so I differentiate these companies into two segments: the sharing economy and the on-demand economy.

The sharing economy is composed of companies having business models based on real sharing. These companies dive below the depths of the corporate ocean to the social ecosystem reef, to access a new supply of physical and human capital in the form of underutilized or latent assets, goods, and expertise. I classify these companies according to three main categories, which are broken down further into two subverticals each:

- **Asset sharing:** personal asset-sharing and corporate asset-sharing.
- **Goods marketplace:** closet sharing and general goods.
- **Specialized services:** professional services and trades services.

The on-demand economy is composed of companies with business models based on the provision of on-demand delivery services or commodity services. Although these companies are also tapping into a new supply of underutilized human capital, I question the sustainability of this business model given rising concerns that these platforms are more exploitative than they are empowering and more focused on delivering convenience than fidelity to customers. Since unskilled labor is a commodity, there is the risk that classifying people as contractors rather than employees could lead to a race to the bottom, as there are no legal requirements for minimum wage, job security, insurance, and benefits. I classify these companies into two main categories, which are broken down further into two subverticals each:

- **Delivery services:** private drivers and goods delivery.
- **Commodity services:** domestic services and nondomestic services.

Extracting Value through Structural Capital

In his book *The Long Tail*, Chris Anderson discusses how the democratization of content will lead to a reversal in the economics of the mass market era. This captures the essence of the abundance economy supply side. But because the sharing/on-demand economy revolves around the democratization of assets, goods, and services, we are dealing not just with falling distribution costs but also with falling structural and customer capital costs. And the underutilized personal assets, goods, and services (which form the long tail that Anderson speaks of as having been "previously dismissed as beneath the economic fringe") can now be exchanged more efficiently and effectively between people below the depths of the corporate ocean.[61] As the sharing economy movement, which began with the idealistic and socially conscious millennial generation, shifts intercohort to baby boomers and Generation X, we are seeing the emergence of a long tail in personal assets, goods, and services, leading to an abundance of supply.

Both sharing economy and on-demand economy companies achieve value extraction by providing sellers with structural capital through two-sided social platforms in creating the long tail of assets, goods, and service markets in these ways:

- **Democratizing the tools of production.** Because individuals can now supply personal assets, goods, and services, the available universe of supply is growing faster than ever.
- **Democratizing the tools of distribution.** The companies are aggregators that lower barriers to entry by providing structural and customer capital, and increase the liquidity of the market in the tail.
- **Connecting supply and demand.** By capitalizing on the filtering efficiency of social networks and facilitating trust, the companies lower the search costs for buyers looking for niche assets, goods, and services.

Capturing Value through Customer Capital

The democratization of content created a long tail and an abundance of supply in digital goods. However, as anyone who has ever written a blog or posted a video on YouTube knows, these non-rival goods are difficult to monetize. Even LinkedIn is still in the early days of monetizing its online marketplace, whereas the sharing/on-demand companies have been able to generate cash from day one.

LinkedIn took the strategic approach to first focus its capital and efforts on building a platform to empower its members to connect with each other and build bonding and bridging capital. Through this "power of we," the company created an appreciating structural asset and a two-sided marketplace that it is just starting to monetize by creating linking capital between the two sides through premium subscription models. The sharing/on-demand economy companies, on the other hand, excel in capturing value, as they offer rival goods in the form of assets, goods, and services through on-demand and transaction-centric business models. They provide sellers with customer capital through their technological infrastructure, which encompasses back-office functions such as social identification (i.e., background checks), payment system, online and live support, tax payments (i.e., issue 1099 income forms in the US or T4A forms in Canada), and return and refund policies.

These companies have focused their efforts on harnessing the power of technology to create social-sharing platforms that facilitate trust by leveraging existing social networks (mainly Facebook) and creating transparency and dual accountability. This enables individuals to form weak ties (i.e., bridging capital) with one another, leading to the personal sharing of goods and services (i.e., linking capital). Because they are focused on transactions, not relationship building, they have been able to generate revenue from day one. The majority of the companies generate revenue by charging a direct commission, which according to my calculations averages just over 20%. The other revenue generation models are based on delivery fees, monthly subscription fees, transaction spreads, and membership fees.

Where the Long Tail Meets the Blue Ocean

In their book *Blue Ocean Strategy: How to Create Uncontested Marketplaces and Make Competition Irrelevant,* W. Chan Kim and Renée Mauborgne advise that "instead of focusing on the competition, you focus on making the competition irrelevant by creating a leap in value for buyers and your company, thereby opening up new and uncontested market space."[62] This is the strategy being pursued by many companies as they create a leap in value for buyers through leveraging leading-edge technology in order to offer unique, authentic, and personal experiences focused on a social mission. In doing so, companies gain access to new tiers of noncustomers, which opens up a new market of nonconsumption, leading to an abundance of demand. The reason abundance economy companies are attractive from an investment perspective is that in addition to accessing nonconsumption on the demand side, they access nonprovision on the supply side. Put another way, the abundance economy is where the long tail meets the blue ocean.

In the paper *Foundations for Growth: How to Identify and Build Disruptive New Businesses*, Clayton Christensen and colleagues recommend that people search for ways to compete against nonconsumption (people's inability to use available products or services because they are too expensive or complicated).[63] To gain further insight into how the abundance economy is achieving disruption, I apply Christensen's three litmus tests to both the supply and the demand sides of the equation:

1. **Does the innovation target customers who in the past haven't been able to "do it for lack of money or skills"?** On the supply side, it was previously too expensive and complex for people to rent out their personal assets (house, car, boat), sell their used goods (clothes), or provide private services (chauffeur, chef). On the demand side, the majority of individuals were not wealthy enough to buy assets for their leisure, refresh their wardrobes monthly, or hire a private driver or chef.
2. **Is the innovation aimed at customers who will wel-**

come a simple product? The high convenience of the leading-edge technology companies is appealing to both sellers and buyers on their platforms and acts as a social accelerator.

3. **Will the innovation help customers do more easily and effectively what they are already trying to do?** The majority of the companies are highly compatible, with established selling and buying norms. As a result, it likely won't take long for the abundance economy to hit a strategic inflection point.

Interestingly, in December 2015, over a year after I came up with the concept of applying Christensen's three litmus tests to the sharing economy, he coauthored the highly critical *Harvard Business Review* article "What Is Disruptive Innovation?" In it, Christensen and colleagues argue that Uber is not a disruptive innovation because it didn't originate in low-end or new-market footholds, and disruptive innovations don't catch on with mainstream customers until quality catches up to their standards.[64] Although I have a high level of respect for Christensen's research, there is a major flaw in this particular thesis, as he fails to look beyond the customer perspective to how marketplace companies like Uber are disrupting from not just the demand side but also the supply side.

III

The Next Generation of Business Strategy

9

The Sharing/On-Demand Economy

The Four Classes of Long Tails

The sharing/on-demand economy companies provide us with examples and insights into the four classes of long tails:

- **Assets:** asset sharing.
- **Goods:** goods marketplace.
- **Expertise and skills:** specialized services.
- **On-demand unskilled labor:** delivery services and commodity services.

Long Tail of Assets: Asset Sharing

As Chris Brogan and Julien Smith proclaim in *The Impact Equation*, "Ideas can help people change the world, and now anyone can become powerful enough to be a catalyst for what matters to them."[65] What I find so inspiring is that in November 2012, only a month after publishing the book, Smith cofounded Breather, an office space–sharing company to "create peaceful, practical, and beautifully curated spaces for people to work, meet, and relax."

Asset sharing is one of the key catalysts leading to a structural shift in society, from one driven by scarcity to one driven by abundance. The asset-sharing companies enable sharing of personal assets (homes, cars, boats) and corporate assets (office space, parking spots, commercial retail space).

The asset-sharing companies create abundance in supply by unbundling assets through a combination of three dimensions—time, space, and use:

- **Time.** All the companies create abundance by allowing individuals or businesses to take assets they own or have leased and share them with others by unbundling them by time (hourly, daily, weekly, monthly). Interestingly, most of the personal assets are available on a daily basis, whereas corporate assets are most commonly available on an hourly basis. The exception in personal asset sharing is HomeAway, which rents vacation homes on a weekly basis. LiquidSpace and PivotDesk offer monthly leases, and members of ClassPass pay on a monthly subscription basis.
- **Space.** The second most popular dimension for unbundling assets is space. For example, Airbnb allows people the opportunity to host guests in their residence by offering them either a private or a shared room. And two corporate asset–sharing companies, LiquidSpace and PeerSpace, allow businesses to subdivide their space by portioning off square footage; PivotDesk promotes cosharing of workspaces. Also exciting is the growth potential of emerging corporate asset–sharing companies that are going beyond the sharing of physical space—for instance, ClassPass is building a long tail in fitness by aggregating the excess capacity of fitness classes in the highly fragmented studio market and selling access via a monthly subscription.

- **Use.** The third dimension of unbundling is use. While this is the least common unbundling method, this innovative repurposing of assets holds the greatest promise for unleashing abundance and maximizing asset utilization. For example, PeerSpace offers individuals and businesses access to unique venues for offsites, events, and media production.

Through unbundling personal assets in terms of time, space, and use, the personal asset–sharing companies are building long tails of supply in inventory of homes and vehicles. Airbnb is in the nascent stages of accessing the potential long tail of 133 million personal homes in the United States alone, which is more than 27 times the 4.9 million guestrooms that make up the head—or central part—of the distribution curve. Likewise, the vehicle-sharing space has similar tail and head dynamics. Given that the average vehicle sits parked for 95% of its life, there is the potential for a very long tail as owners start to view their vehicles as an easy source of passive income. The current 2.1 million rental vehicles in the corporate ocean account for less than 1% of the 258 million registered vehicles.

At the same time, a structural shift from ownership to access, combined with growth in carpooling services such as Lyft Line, UberPool, and Uber's Driver Destinations, would reduce future new vehicle demand, negatively impacting sales of new vehicles. However, this is not starting to happen yet. The number of new cars and light trucks sold in the United States in 2014 actually rose by 5.9% to 16.5 million, with the average price increasing 4.5% to over US$32,000, resulting in total new vehicle sales climbing by over 10% to US$535 billion. And in 2015, sales of new cars and light trucks rose by a further 5.7% to 17.5 million, with the average price increasing 2.5% to over US$33,000, resulting in total new vehicle sales climbing by a further 8.3% to US$580 billion.

An interesting twist on corporate asset unbundling is WeWork. Launched in 2010, WeWork is revolutionizing traditional office space by building a long tail in coworking spaces through the signing of long-term leases on large areas of office space and then converting them into incubator-like spaces with a communal atmosphere and hip ambience. Although WeWork does not own the assets, it does control the supply side by taking on long-term liabilities in the form of property leases, so it does not meet our criteria for a two-sided marketplace. Despite this, the company has been able to scale quickly, and investors seem to like its innovative business model—it is valued at US$16 billion as of August 2016.

Long Tail of Goods: Goods Marketplace

The goods marketplace companies create abundance through leveraging advances in technology (e.g., smartphone cameras, apps, social networks, user-generated reviews, cashless payment systems) to provide liquidity for the online exchange of secondhand goods. This new form of online exchange is especially appealing to the millennial cohort of cost-conscious and socially conscious consumers. The closet-sharing companies are creating a long tail inventory in women's secondhand apparel, while the general goods companies are creating a long tail in niche secondhand items ranging from electronics to art to furniture to general goods.

Long Tail of Expertise and Skills: Specialized Services

The challenge many nontraditional workers (freelancers, consultants, contractors, artisans) face is how to get paid for their expertise and skills—though they can originate value through human and intellectual capital, they lack the structural capital to extract that value (turn their talents into actions) and the customer capital to capture that value (turn their actions into money). As a result, many people with specialized skills still need to be traditionally employed, either by a small or medium-sized enterprise (SME) or by a large corporation. The wonder of specialized services companies, which harness the power of technology to create a two-sided social-sharing platform, is that they empower people with proprietary skills to independently monetize their expertise by providing them with both structural and customer capital.

Specialized services companies create long tails of experts operating over a wide range of professions and trades. One of the most successful start-ups is Thumbtack, which is transacting over 5 million projects per year and bringing more than US$1 billion in business from its base of some 230,000 active professionals. When I was in San Francisco in October 2015, I had the privilege to meet with one of Thumbtack's cofounders, Sander Daniels, who launched the company in August 2008 to help "you accomplish the personal projects that are central to your life."

The biggest growth potential will come from specialized services companies that are able to create blue oceans by partnering with professional or trade associations to unbundle their members' talent in terms of both time and space. The most attractive professions and trades are those with the largest value gap between what wage or salary people can command in the corporate marketplace and what they can charge on their own. One of the professions with the largest value gap is business consulting. This is the space being tackled by HourlyNerd (renamed Catalant in July 2016), cofounded as part of a class project back in February 2013 by Rob Biederman and three of his Harvard MBA classmates, with the mission to "disrupt the 100-year-old consulting industry by connecting businesses to MBA alumni."

On the trades side, there is a large value gap for chefs. According to the US Bureau of Labor Statistics, just over 100,000 chefs and head cooks in the United States make an average of US$20 per hour, equating to US$46,000 a year. We could see an exodus of chefs from restaurants if there were a platform that enabled them to generate over US$46,000 in profits, with the added advantage of being able to control their own schedules and enjoy the variety and emotional benefits of cooking for people in their homes, versus working in an institutional kitchen. However, this utopian world might not be realistic: both Kitchit and Kitchensurfing decided in 2015 to pivot to the more exploitative on-demand chef model and ended up filing for bankruptcy in April 2016. This leaves EatWith, which has stayed true to its original vision of creating a long tail of chefs to host community dinners at their homes or cater private events.

I am excited about the emergence of specialized services companies, which fulfill the unmet desire for people to control their destiny by being their own boss. I just hope these companies aren't tempted to pivot to the less empowering on-demand talent model.

Long Tail of Unskilled Labor: Delivery Services and Commodity Services

Is the long tail business model of on-demand economy companies sustainable? These companies have been able to quickly scale their supply base by creating opportunities for people to easily monetize their underutilized human capital by performing gigs such as driving, delivery, caregiving, and housecleaning. Such platforms are empowering in the sense that they allow people the freedom to work where and when they want, using their own assets (e.g., car, bike, tools). However, unlike the specialized services workers, the delivery services workers and commodity services workers do not have differentiable expertise or skills. Having no protections in place for the unskilled workers could lead to a slippery downward slope to the lowest common denominator, due to lack of minimum wage, job security, insurance, and benefits. Consequently, we are starting to see an increased level of scrutiny from government, regulators, and the workers themselves as they start to question the potential exploitative power of companies building supply bases on the backs of unskilled human capital.

In light of this rising legal and regulatory risk of worker misclassification, on-demand companies have proactively announced intentions to shift their supply base from contractors to employees. For example, on July 1, 2015, Shyp announced it would be switching the status of its couriers from independent contractors to part-time or full-time employees. Although this will negatively impact the company's cost structure by increasing its labor costs by 30% and shifting them from variable to fixed, as the company's CEO, Kevin Gibbon, wrote in his blog post, "this move is an investment in a longer-term relationship with our couriers, which we believe will ultimately create the best experience for our customers."

Two weeks later, Instacart announced it would be offering its contract "personal shoppers" the option to apply for employee status, though as of August 2016 it was not offering the same alternative for its delivery drivers. And two weeks after that, Luxe announced it would be switching all its valets from contractor to employee status effective immediately and would pay for all work-related expenses, such as phone usage, jackets, workers' compensation, overtime, unemployment insurance, Medicare and Social Security taxes, and health benefits (for full-time employees). On his blog, Curtis Lee, CEO and founder of Luxe, wrote: "As we grow, we have realized the need to assert more direct control over the customer experience and provide our valets with career development opportunities and benefits, none of which are possible within the boundaries of the 1099 model." And at the beginning of August, Sprig announced it would be transitioning its servers from contractors to employees and would be offering them the opportunity to earn stock option grants. According to the *Inc.* article "Yet Another On-Demand Startup Exits the Gig Economy," Sprig's CEO, Gagan Biyani, hopes to recoup some of the increased costs through increased productivity, improved employee retention, and improved relationships with customers, as it will be able to train and uniform its employees.[66]

It is only a matter of time before regulations force the majority of the companies in the on-demand economy to shift the status of their supply base from independent contractors to employees. From an abundance perspective, this would eliminate these companies' long tail advantage and impair their ability for accelerated value creation, since they would no longer be able to scale their supply side, with no time or capital constraints.

In terms of the impact on costs, in addition to an estimated 20% to 30% escalation in labor costs, their cost structure would shift from variable to fixed, and they would no longer have an asset-light business model. The rapid proliferation in on-demand companies reminds me of Webvan, the online grocery delivery business founded in late 1999, which went bankrupt in 2001 during the largest dot-com bust in history. Ironically, the key difference is that the on-demand companies have asset-light business models compared with Webvan's heavy upfront infrastructure costs (warehouses, vans, employees). But this advantage would disappear with the elimination of their long tail.

Delivery Services

The delivery services category is composed of personal driver companies and goods delivery companies. As noted earlier, Uber is not just creating a long tail—the company is setting out to completely destroy the highly fragmented, privately owned taxi companies swimming in the corporate ocean. As evidence of Uber's ambitions to leverage its long tail of supply and structural database to evolve into a complex transportation logistics company, on August 17, 2015, it changed its app interface to separate "rides" from "eats." Even though as of August 2016, UberEATS is available in only 15 US cities, 3 Canadian cities, and 4 international cities (London, Paris, Melbourne, Sydney), this emphasis on food delivery indicates the emergence of a second key category. In addition to posing a direct competitive threat to the goods delivery companies, Uber poses a potential threat to the dominant incumbents, UPS and Federal Express. UPS appears to be aware of this threat. As I detail further in Chapter

10, UPS strategically invested in Deliv, an on-demand goods delivery company, in February 2016.

The question of whether drivers should be classified as independent contractors or employees is highly controversial. On one hand, the taxi industry sets precedence for private-driver companies such as Uber, since most taxi drivers are deemed independent contractors in that they lease the vehicles from the taxi company and pay for all associated costs, such as gas and maintenance. Uber could argue it provides even more freedom for its drivers, as they are free to set their own timetables and drive their own vehicles, and are able to drive for other companies. However, in a nonprecedent-setting ruling in early June 2015, the California Labor Commissioner's Office ruled that Uber's drivers should be classified as employees because Uber is not just a neutral technology platform that acts as a matchmaker between riders and drivers—it provides drivers with phones, sets prices for the fares, vets the drivers and their vehicles, can fire drivers if their rating falls below 4.6 out of 5.0 stars, and can deactivate their app if inactive for more than 180 days. In April 2016, Uber announced it had reached a settlement for two class-action lawsuits (in California and Massachusetts) and agreed to pay concessions of up to US$100 million to 385,000 drivers in exchange for being able to maintain the status of its drivers as independent contractors. However, less than a month later, a nationwide class-action lawsuit (excluding the states of California and Massachusetts) was filed against the company.

The goods delivery vertical is facing the most immediate legal and regulatory threat of worker misclassification. The companies that will be able to preserve their long tail advantage are those with B2B (business-to-business) strategies like Drizly, which is building its long tail in alcohol delivery by using technology to aggregate retailers of beer, wine, and spirits and then leveraging the retailers' own employee base to do the delivery. Although personal driver companies like Uber and Lyft are on-demand delivery services companies built on a long tail of unskilled human capital, it could be argued that drivers who use it as a carpooling tool on their commute to and from work, or to run errands, are not employees. A positive case could also be made for UberPool and Lyft Line, which facilitate car sharing. Two others that leverage excess capacity are B2B companies such as Cargomatic, which connects shippers with local truckers with excess capacity, and P2P (peer-to-peer) companies such as Roadie, a neighbor-to-neighbor shipping network, which is like carpooling for cargo.

Commodity Services

The commodity services category comprises domestic services and nondomestic services companies. I question the overall sustainability of the business models of the commodity services companies in light of the bankruptcy of home-cleaning company Homejoy, which closed its doors in July 2015. In addition to red oceans of increasing competition, some of these companies are starting to face lawsuits based on the potential misclassification of their workers. Further, domestic services companies that provide scheduled personal routine services like housecleaning, childcare, and eldercare have been traditionally built on long-term, trust-based relationships, so they could face challenges in building demand and may face attrition down the road.

Although Luxe is the only company in the commodity services vertical that has so far elected to shift its supply base over to employees, it is only a matter of time before the remaining companies will be forced to follow suit. The two commodity services companies that should be able to preserve their long tail advantage are asset-sharing hybrids like DogVacay and Rover.com. Both are like an Airbnb for dogs, with the added benefit of personal pet care at a private home. The other exception is Bellhops, launched by Cameron Doody in October 2012 to "make moving affordable and easy for everyone." Given that in November 2015 Bellhops was ranked number four on *Entrepreneur* magazine's "25 Best Large-Company Cultures" list and that it provides a very appealing opportunity for college students to earn income between their studies, the company is likely in a favorable position to continue growing its long tail.

To Create a Movement, Be a Rebel with a Cause

Speaking in terms of Joseph Campbell archetypes in *The Hero's Journey*, Travis Kalanick and Brian Chesky are joining the league of tricksters with founders like Chip Wilson, Howard Schultz, Steve Ells, John Mackey, Richard Barton, and Reid Hoffman (founder of LinkedIn)—they all relish disruption of the status quo and love to turn the Ordinary World into chaos. In my research report *The New Era of Economic Abundance*, published in September 2015, I counted that 26 of the 75 companies I profiled, or just over a third of the sharing/on-demand economy universe, are composed of such rebels with a cause. As discussed in Chapter 3, these companies are not just online marketplaces but movements that promote accessibility, sustainability, and community.

LinkedIn and Zillow have built successful movements based on empowerment through transparency and connections via their accessibility-focused social missions to "create economic opportunity for every professional in the world" and to "lead a revolution in online real estate to empower consumers" respectively. Interestingly, half of the social mission–oriented, sharing economy companies promote accessibility, with a focus on transportation, education, and health care. On the transportation front, Uber's mission is to "bring transportation as reliable as running water to everyone, everywhere," while Turo's mission is to "connect vehicle owners whose cars would otherwise be idle with people who need a car." In the education space, Udemy's mission is to "help anyone learn anything," Skillshare's is to "provide universal access to high-quality learning," Codecademy's is to "teach the world to program," and Duolingo's is to "translate the web into every language and make language education accessible to the masses." On the healthcare front, ZocDoc's mission is to "improve the accessibility of heath care" while Sherpaa's is to "deliver care more effectively and less expensively." There are also companies promoting accessibility to art (Artsy), farm-to-neighbor access to fresh food (Farmigo), and the love of a dog (Rover.com). And lastly, that of Quirky (which filed for bankruptcy in September 2015) was to "make invention accessible," while that of Storefront (which shut down in March 2016 but then merged with Oui Open in September 2016) was to "make retail space more accessible, empower store owners/merchants, and foster local economies."

Whole Foods Market and Chipotle Mexican Grill were incredibly successful in building a cult-like following with their sustainability-focused missions to "help support the health, well-being, and healing of other people and the planet" and of "food with integrity." Likewise, sustainability is a passion for eight of the sharing companies with a focus on taking cars off the road, reducing wasteful consumption, and promoting local commerce. On the car front, Lyft's mission is to "take cars off the road, not replacing or augmenting existing systems," while Getaround's is to "have a world with fewer cars, without traffic jams, and less pollution," That of FlightCar (which shut down in July 2016) was to "reduce the number of parked cars sitting idle at airports by connecting them with travelers that need cars." On the consumption front, Yerdle's mission is to "reduce the durable consumer goods we all need to buy by 25%," Listia's is to "create a universal currency to reward people for giving," and Pley's is to "raise a more creative and skillful generation, emphasizing sharing, reducing waste, and giving back." On the local commerce front, Etsy's mission is to "reimagine commerce in ways that build a more fulfilling and lasting world," while Good Eggs' is to "grow and sustain local food systems worldwide."

lululemon and Starbucks succeeded in building strong cult-like movements based on their strong community focus, which brought to life their greater purpose to "create components for people to live a longer, healthier, and more fun life" and to "inspire and nurture the human spirit—one person, one cup and one neighborhood at a time" respectively. In my research report, I identified five sharing economy companies looking to bring people together through their platforms. In terms of asset sharing, Airbnb's mission is to "imagine a world where you can belong anywhere," while PivotDesk's is to "build a marketplace for office sharing." VarageSale's mission is to "create a safe place for people to sell old treasures and find new ones and swap stories and stuff," and EatWith's is to "create a new dining experience." And that of Sidecar (which shut down in December 2015) was to "build the largest social transportation network in the world."

As evidenced by the recent shutdowns of Quirky, Sidecar, and FlightCar, a start-up's social mission can be the source of its downfall if the founder has not developed a sustainable business model to support his or her aspirational ambitions.

"Halo Polishing" is the New Greenwashing

In retrospect, I was a bit Pollyanna-ish in my thinking that a greater purpose would create a strong stakeholder foundation. Although I do believe advocacy is an important social value driver, I'm concerned about the increasing number of companies hopping on the "social mission" and "greater purpose" bandwagons. This trend was parodied in the popular TV series *Silicon Valley* in which the founder of Hooli, Gavin Belson, preaches, "Hooli is about innovative technology that makes a difference, transforming the world as we know it, making the world a better place." Ironically, in April 2016, Dan Lyons, a former technology editor for *Newsweek* and coproducer and writer for *Silicon Valley* published a brilliant exposé on his experience of working at Hubspot, titled *Disrupted: My Misadventure in the Start-Up Bubble*, in which he astutely and critically observes:[67]

> *On top of the fun stuff you create a mythology that attempts to make the work seem meaningful. Supposedly millennials don't care so much about money, but they're very motivated by a sense of mission. So, you give them a mission... You make up a culture code and talk about creating a company that everyone can love. You dangle the prospect that some might get rich.*

I agree with Lyons. Just as the rising popularity of corporate social responsibility over the last decade led companies to engage in greenwashing, the rising popularity of having a social mission is leading some companies to engage in what I call "halo polishing." As Douglas Holt and Douglas Cameron wisely state in *Cultural Strategy*, "Companies that walk the walk, living their ideology every day in their business, have much more credibility with today's consumers than do companies that promote brands as champions of an ideology that is unrelated to the company's business practices."[68]

I'm worried that some of these companies are just pretending to care about customers, employees, and suppliers in order to attract the spending dollars and talents of the socially conscious generation of millennials. Even more concerning to me is the start-up companies claiming to have a social mission, hoping to detract investors from unrealistic revenue projections, weak competitive positions, and unsustainable business models. But once the capital stops flowing, this will be exposed. As Dave Logan, John King, and Halee Fischer-Wright observe in their book *Tribal Leadership*, "In the early 2000s, we saw an aberration: Stage Five cultures with no discernable business expertise. Many weren't making money but did have large infusions of investment capital. Over the long term, culture and strategic performance correlate, with the higher factor falling to the lower. Thus, a company with a great culture and low strategic performance will, over time, find that its culture erodes: good people leave."[69]

Although many of these companies may preach strong values and concern for customers, employees, and suppliers, ultimately it comes down to how these groups are treated. And although the threat of lawsuits may prevent some people from sharing negative experiences, social media is an incredibly empowering force that gives everyone a voice, and so the truth will eventually come out. When this happens, such companies will discover the fragility of their foundations.

The Economics behind Long Tails and Blue Oceans

Based on my research into the economic characteristics of the top 75 North American sharing/on-demand economy companies, I uncovered the interesting truth that it is harder for most start-ups to build the supply side (long tails) than the demand side (blue oceans).

The Economics behind Long Tails

Although the majority of the top companies have high supply compatibility, over three-quarters of them face challenges resulting from having a high marginal cost of supply, low marginal utility of supply, low market liquidity, and no corporate exposure. What's interesting is that the verticals that are becoming the most competitive (e.g., closet sharing, goods delivery, domestic services, nondomestic services) tend to be the ones with the least favorable supply-side characteristics. And not surprisingly, the companies with the most favorable supply-side characteristics are based on the sharing of personal assets, corporate assets, or professional expertise. To provide greater insight into why economics matter, let's take a look at Airbnb, one of the few start-ups with highly attractive supply-side characteristics

Airbnb has a highly enviable close-to-zero marginal cost of supply, as the depreciation of underutilized homes is minimal and the rental process requires only a modest amount of time and effort. From an economic perspective, this low marginal cost of supply translates into low price elasticity of supply, resulting in a rare flat supply curve. This is important because it implies that people will still use the platform to rent out their place even if increased supply leads to lower pricing. It has a high level of market liquidity, since although it rents out physical assets (homes), it operates in the travel marketplace and has been able to leverage this cross-platform dynamic to quickly expand to 190 countries, and now has the opportunity to expand into new cities and towns within the countries themselves. As a personal asset–sharing platform, one of Airbnb's greatest challenges (and opportunities) is supply compatibility: although people may be comfortable renting out their vacation homes, they might not want to deal with the personal space and privacy issues of renting out private or shared rooms in their primary residence, or trust strangers with it. Lack of financial motivation can also be seen as a hurdle, as affluent homeowners might not desire any additional income.

Airbnb is also well positioned to increase its average units per seller—these sellers increasingly utilize Airbnb to generate passive income on top of the structural and customer capital provided by its platform. As corporations have more resources and greater ability to scale, adding them to the supply mix will increase Airbnb's average number of accommodation listings per seller. Airbnb has an extremely high marginal utility of supply, since its sellers on average currently rent out their home for only two months of the year. As its sellers move up the experience curve and gain more positive reviews and get better at marketing, they will likely increase the number of nights per year or even carve out more rooms in their house to rent. This is a huge growth lever: if Airbnb's sellers added on average just three incremental nights per year, its inventory of available room nights would increase by 3.6 million.

When you think about it, established companies have a competitive advantage over start-ups when it comes to building a long tail of inventory. They are better positioned to originate value because they can access their own inventory of physical or human capital, or collaborate with one of their stakeholder groups to access their underutilized capital. And they can improve the value-extraction process by leveraging both their tangible assets (e.g., property, plant, and equipment) and intangible assets (e.g., brand equity, proprietary data) to build a marketplace driven by SaaS (software-as-a-service) that improves the tools of production for their stakeholders to supply their assets, goods, skills and expertise, and time. This will enable companies to cost-effectively accelerate their inventory growth by minimizing supplier acquisition costs while reducing the rate of supplier attrition.

The Economics behind Blue Oceans

As mentioned, it is easier for companies to build the demand side. The majority of the sharing/on-demand economy companies have high-demand compatibility, whereas only half of them face challenges resulting from having low fidelity of demand and no corporate exposure, and less than half have low marginal utility of demand. The vertical with the most favorable demand-side characteristic is professional services, followed by corporate asset sharing, private drivers, and goods delivery. The verticals with the least favorable demand-side characteristics are domestic services, nondomestic services, closet sharing, and general goods.

Existing companies have a strong competitive advantage over start-ups when it comes to building their blue ocean strategy: customer capital. But as Mitch Joel advises in *Ctrl Alt Delete*, "The new consumers are not linear. They are scattered. They are squiggly. They are connected—not only to one another, but also to the world—and their connectivity and engagement are highly untethered."[70] In crafting their strategy, companies would be wise to heed the advice of Saul J. Berman in his book *Not for Free*:[71]

> *Your existing customers, for better or for worse, will drive how you pursue new revenue streams. Outdated ideas of who your customers are cause blunders when their behaviors and attitudes shift. Do not just look at what your consumer segments are doing now but what they want to be doing, or will be doing, in the future. Do not just look at where you operate in your value chain but where value might be shifting and what you would have to do to play there as well.*

By crafting a blue ocean strategy for a new product or service with a compelling customer value proposition that leverages their existing customer base, companies will be able to accelerate their revenue growth, minimize their customer acquisition costs, and reduce their rate of customer attrition. Again, one of the few start-ups with highly favorable demand characteristics is Airbnb. Although people are comfortable using online travel accommodation sites like Expedia, Kayak, Booking.com, and Priceline to book hotels and vacation rentals, many are still not comfortable with the thought of staying in a stranger's home, especially in a private or shared room. This makes demand compatibility a massive growth lever for Airbnb in terms of attracting customers to its platform. By providing a high-fidelity experience for its customers, Airbnb should be able to sustain its blue ocean and continue to attract tiers of noncustomers through a positive social network effect. Airbnb offers a superior value proposition for its customers in terms of:

- **Functional.** Airbnb offers excellent value for money, convenience with on-demand booking options, and an extremely wide variety of choice with the option to rent an entire home, a private room, or a shared room in cities and towns in 190 countries.
- **Emotional.** The company provides a highly unique asset, a very personalized experience, and a potential emotional bonding opportunity with the host.
- **Psychological.** Airbnb is advancing its community-based social mission by building a collaborative community around its new symbol, the Bélo, and is encouraging both its hosts and guests to design their own "bélo" and share their stories.

Airbnb is also well positioned to increase its average spend per buyer by increasing the average transaction price and the frequency of transactions. One of the company's significant growth opportunities is to advance from leisure to business travelers. The US$300 billion US business travel market is especially attractive: business travel only accounts for 10% of Airbnb's revenue. The company is looking to aggressively expand its current corporate account base from 250 companies through its newly revamped Airbnb for Business platform, which offers complimentary specialized features such as a dashboard that keeps track of employee spending, as well as centralized billing. Airbnb is positioned to increase the number of transactions by gaining a greater share of the travel accommodation budget of travelers as they gain more confidence in Airbnb's platform and its opportunity to offer them unique, high-fidelity travel experiences.

10

The New White Space

Ford No Longer Sees Its Future as Just an Auto Company

> *Strategic inflection point is the moment at which the balance of forces shifts from the old structure, from the old ways of competing, to the new.*

—Richard S. Tedlow, *Denial*

On January 11, 2016, at the North American International Auto Show in Detroit, Mark Fields, CEO of Ford Motor Company, declared that he no longer sees its future as just an auto company but as an auto and mobility company. And with this, he announced the upcoming launch in April of FordPass, Ford's new digital and physical marketplace platform, driven by the company's desire to create true, authentic long-term relationships with its customers and continue to improve its products and services.

When I listened to Fields discuss how FordPass was inspired by the company's desire to emulate the successful customer-centric strategy of companies like Apple, Nike, Nespresso, Ikea, and Disney, I couldn't help but smile. At last companies were starting to recognize the power of the new social value driver of advocacy.

Interestingly, Ford's announcement came only weeks after GM announced it would be making a US$500 million strategic investment in Lyft, and a week prior to GM's acquisition of recently shuttered car-sharing platform Sidecar. Back on November 16, 2015, when I published the article "A Wake-Up Call for CEOs—from Amazon, LinkedIn, and Expedia," discussing how those three tech companies had made recent strategic moves into the sharing/on-demand economy, I never imagined that they would be joined so quickly by century-old Industrial Age stalwarts like GM and Ford. This is making me even more confident in my prediction that over the next decade we will see the leading companies from every industry sector embrace some form of ubernomics as they seek to shift their business model from scarcity to abundance.

The analyst in me is curious to watch Ford's transformation from an auto company into an auto and mobility company. For example, in the United States alone, Ford sold 2.6 million vehicles in 2015 at an average sales price of US$35,000, generating US$91.6 billion of revenue, which, based on its 10.2% operating margin, yielded US$9.3 billion of pretax income. So I am guessing that Ford is not going to want to sacrifice this product cash cow anytime soon. But regardless, Fields and his team should be commended for being proactive in addressing the recent convergence of structural shifts in the following three forces:

- **Technological.** FordPass is a digital platform that leverages recent advances in technology such as smartphones, mobile apps, cashless payment systems, and online marketplaces.
- **Economical.** FordPass recognizes that millennials value access over ownership.
- **Societal.** FordPass embodies Ford's new accessibility-

focused social mission to "rethink the way you move."

Through FordPass, Ford is looking to meet the evolving needs and desires of the uberized consumer by advancing its consumer value proposition in terms of:

- **Functional.** FordPass offers superior convenience, as you can use the mobile app to remotely start your car, connect 24/7 with a live personal mobility assistant through FordGuides, and locate and pay for parking through FordPay.
- **Emotional.** FordPass will help Ford strengthen the emotional connection its customers have with the Ford brand through FordGuides and FordHubs, its new retail storefront innovation hubs.
- **Psychological.** FordPass will help Ford create a psychological attachment with its customers and create a movement through its new accessibility-focused social mission to "rethink the way you move."

Under the new laws of ubernomics, the following economic forces will act as growth catalysts for Ford:

Long tail of supply. Although Ford is not building its own long tail of supply, it is accessing a long tail of inventory through numerous initiatives:

- Ford is looking to build a long tail consisting of its fleet of shuttle vans through its Dynamic Shuttle service, which it is currently beta testing with its employees.
- Ford is looking to build a long tail of its vehicles for share through a shared-leasing program that it is currently pilot testing in Austin.
- Ford was looking to partner with FlightCar to access the

long tail of vehicles left at airports by travelers. However, in July 2016, FlightCar unexpectedly shut down and sold its technology platform to Mercedes-Benz's North American Research & Development division.

Blue ocean of demand. Ford is looking to expand its TAM (total addressable market) beyond its existing customer base by tapping into a new pool of former and non-Ford owners looking to access its complimentary, innovative FordPass services, as well as of millennials who are seeking access to vehicles rather than ownership.

***

At the Urban Land Institute conference in Paris in early February 2016, Mark Gilbreath, founder of LiquidSpace, gave a fascinating presentation, which he concluded by proposing the radical concept that we are seeing the inevitable move to the era of Peak Car, which will be followed by the era of Peak Office. His company has created a new legal framework to enable landlords and tenants to unbundle the traditional long-term office lease, and I wonder how long it will be before we see companies like Ford unbundle their vehicle lease into a similar standardized license. If, like me, you have had the unfortunate experience of having gone through the stressful and financial burden of trying to exit from an ironclad vehicle lease agreement, I'm guessing you would also welcome a more innovative and flexible alternative.

When I attended the 2001 North American International Auto Show as an associate analyst working on Deutsche Bank's Global Auto Team in New York City, I never imagined that, 15 years later, Ford would announce it no longer saw its future as just an auto company. I can't help but think about what this means for the future of other companies. As I discuss in Chapters 5 and 9, and what Mark Fields and his team at Ford have discovered, business transformation starts with questioning the status quo and becoming a rebel with a cause.

Hyatt Launches "Airbnb" of Boutique Hotels

In March 2016, Hyatt became the latest company to embrace ubernomics, joining the growing ranks of forward-thinking companies. As president and CEO of Hyatt Hotels Corporation Mark Hoplamazian is quoted as saying in the company's press release, "The Unbound Collection by Hyatt provides us with a myriad of opportunities to grow, not only in new markets, but also in places we know our guests want to go."[72] Instead of acting from a scarcity mindset of acquiring independent boutique hotels, as it may have done in the past, Hyatt is embracing a mindset of abundance by creating a corporate asset–sharing marketplace to collaborate with them.

The new Unbound Collection is a brilliant strategic move, as it will enable Hyatt to leverage its strong brand equity, customer capital, proprietary data, and strength of its marketing and loyalty program. Hyatt generated US$4.3 billion of revenue in 2015, so I am guessing this will not have a material impact on its top line. However, it will enable Hyatt to start defying traditional economic principles of scarcity in terms of both supply and demand:

- **Long tail of supply.** Hyatt's inventory growth for its Unbound Collection is not limited by traditional time or capital constraints, as it is able to access a long tail of underutilized rooms at independent boutique hotels.
- **Blue ocean of demand.** Hyatt's revenue growth is not

constrained by existing demand, as it will be able to market these hotels to millennials as well as to customers who have switched or are looking to switch to Airbnb. This will enable Hyatt to access new blue ocean market demand, expanding its TAM beyond traditional hotel stays.

Hyatt is not new to the sharing economy. In June 2015, the company participated as a strategic investor in the US$40 million Series D round of onefinestay, a London-based personal asset–sharing company for fine homes, founded in 2009 with the mission of "handmade hospitality." And roughly five months later, during Hyatt's 3Q15 analyst conference call, as mentioned earlier, Hoplamazian foreshadowed the company's entry into the sharing economy with these insightful remarks: "From our perspective, we've always—we've, for some time, looked at this whole sharing economy dynamic as a broad consumer issue and the consumer behavioral change and we've always been drawn towards it, not sort of away from it."[73]

Through its experience in working with onefinestay, Hyatt discovered that the convergence of structural shifts in technology, the economy, and society has led to the emergence of the uberized consumer. A lot of people, especially millennials, are no longer content to stay at soulless, cookie-cutter hotels and are looking to personal asset–sharing marketplaces like Airbnb and onefinestay, which offer a wider variety of choice and more of an authentic and emotional travel experience.

It's interesting to contrast Hoplamazian's comments with those made in February 2016 by Christopher Nassetta, president and CEO of Hilton Worldwide, on the company's 4Q15 conference call when he was asked about Airbnb: "I think we believe that these are different businesses. There is overlap in our customer base. We do not see any material impact from it. I think testimonial to that is that the industry is at the highest levels of rates and occupancy that we've ever seen in history."[74] As the sharing/on-demand economy is still in its infancy, I wonder whether most corporate executives share Nassetta's views and are still indifferent to it.

The encouraging news is that more and more companies over a wide range of industries are starting to see ubernomics as an opportunity. In April 2016, only a month after Hyatt's launch of the Unbound Collection, Accor S.A., a Paris-based hotel operator with 3,700 hotels and a market capitalization over US$10 billion, acquired onefinestay for US$170 million. And in February 2016, UPS participated as a strategic investor in the US$28 million Series B round for Deliv, an on-demand goods delivery company founded in January 2012 that is building a long tail in individuals delivering goods, with the objective to "make online shopping simpler and more convenient."

As Rimas Kapeskas, managing director of the UPS Strategic Enterprise Fund, stated in the Reuters article "With Deliv Investment, UPS Hopes to Study Same-Day Delivery Market," "We do not participate in the on-demand business as much, and the consumer side of this is still a bit of a mystery to us… This is a rapidly evolving marketplace and we thought we could learn more by being close to it."[75] By 2017, we could see UPS act upon what it learns and capitalize on its brand equity, logistics infrastructure, proprietary data, and customer capital to either acquire Deliv or build its own marketplace in the goods delivery vertical. It would be a good defensive move against Amazon Flex.

I strongly believe that ubernomics represents the next generation of business strategy, and agree completely with Hoplamazian's words during Hyatt's 3Q15 conference call in November

2015:"I just think that, as we evolve over time, this is going to be a part of how people travel and how people experience the world not just in lodging, but in transportation and travel more broadly. And I think that means we have to go towards it and understand it better and better."Hyatt is a great example of how a company can apply the principles of ubernomics to create a sustainable competitive advantage. As Scott D. Anthony advises in *The Little Black Book of Innovation*, "Success requires waking up every day and realizing that today's sources of competitive advantage will not be tomorrow's."[76]

From a cost of capital perspective, it is an asset-light and low-cost producer strategy that enables Hyatt to source inventory from independent boutique hotels at close-to-zero marginal cost. From a customer perspective, it increases customer loyalty and increases switching costs by meeting the needs and desires of its guests that are seeking a wider variety of choice and more of an authentic and emotional travel experience. From an intangible-asset perspective, it enables Hyatt to create a marketplace that leverages its strong brand equity, customer capital, and proprietary data, as well as the strength of its marketing/loyalty program. And from a growth perspective, it creates a structural asset that appreciates in value as it adds more independent boutique hotels to its collection, attracting new guests, leading to the ultimate network effect.

As Harley Manning and Kerry Bodine state in *Outside In*, "Competitive barriers of the past—manufacturing strength, distribution power, information mastery—can't save companies today. One by one, each of these corporate investments have been commoditized."[77]This is why ubernomics is so powerful: it enables companies to create a sustainable competitive advantage, not through exploitative barriers but by building generative economic moats.

Capital Hacking—Why Start-Ups Need Strategic Investors

I met up for coffee in late March 2016 with my friend Ryan Spong, cofounder and CEO of Foodee, a goods delivery company. While I still feel a certain amount of skepticism about on-demand start-ups (like I said earlier, I see most of them as tech-enabled conveniences that treat human capital as just a commodity), Ryan's company is different. His delivery force is made up of employees, not contractors, and he is building a long tail of underutilized restaurant capacity to bring "the food culture to the office culture" and make high-quality food accessible to time-deprived office workers.

I congratulated Ryan on his Series A round of C$6 million, which is pretty impressive considering venture capital is becoming more scarce. In fact, only 4 of the 75 sharing/on-demand economy companies in my database raised capital in 1Q16, versus 18 companies in 1Q15. As we continued to chat, I mentioned how disheartened I was to discover that Storefront, a commercial retail space, corporate asset–sharing company that had been founded in early 2012 with the social mission to "make retail space more accessible, empower store owners/merchants, and foster local economies" had just closed its doors. With rumors of unicorns at risk of turning into "unicorpses" and "donkeys," I told Ryan I was becoming increasingly convinced that the marketplace model wouldn't translate for most start-ups.

I went on to share with him my belief that the low barriers to entry, the attractive economics, and the recent flood of capital to these start-ups have led to the perfect storm in terms of Porter's Five Forces. I explained that, because of the unique dynamics of a marketplace, the heightened competitive environment is leading to increasing bargaining power for both suppliers and buyers and, as a result, we are starting to see a double jeopardy of rising acquisition costs and increased attrition on both the supply and demand sides.

Ryan told me he had started to put together a list of potential strategic partners for Foodee (drawing upon his former life as a Wall Street investment banker) and asked whether I had developed an ubernomics matrix to help start-ups. I had not, but Ryan inspired me to start doing some research. Upon looking up the CrunchBase investor list for each of the 75 sharing/on-demand economy start-ups in my database, I discovered that just under a quarter had raised capital from strategic investors, from 9 of the 10 verticals. It's interesting because, of the 17 companies with strategic investors, only 4 received just financial capital, 7 received supplier capital also, 5 received customer capital also, and 1 received structural capital also.

Ironically, the supply side represents one of the biggest risks for companies, as they do not own or control the inventory of assets, goods, or services. And many start-ups' business models face challenges in terms of having a high marginal cost of supply, a low marginal utility of demand, and low market liquidity. One of the best supplier capital hacks was by Mark Gilbreath, founder of office workspace–sharing company LiquidSpace, who brought in GPT Group, CBRE Group, and Steelcase as investors in early 2013. Also worth noting is that GM, which recently invested in Lyft, was actually an early strategic investor in RelayRides (now Turo), back in late 2011. But companies need to do their due diligence, as GE learned—the hard way. According to the *Wall Street Journal* article "GE Says Quirky Has Hurt Its Reputation," GE claimed in Quirky's December 2015 bankruptcy-related documents that Quirky "caused substantial damage to the reputation of GE and to its trademark."[78]

Although the verticals all started out as blue oceans of demand, many of them are starting to turn blood red as more players enter the waters. With customers becoming more difficult and costly to acquire, this increase in players is leading to demand risk, since start-ups have to build their customer relationships from scratch. And many start-ups' business models face challenges in terms of having a low marginal utility of demand, low fidelity of demand, and the absence of corporate buyer exposure.

Look, for example, at Whole Foods Market. In September 2014, it partnered with Instacart to "offer its customers the convenience of delivery without having to handle the logistics themselves" and then made an undisclosed strategic investment in the company in February 2016.[79] Another customer capital hack strategy for start-ups is to seek financing from their own customers, as did Getable (with four of its construction companies) and HourlyNerd (GE Ventures). Or they can seek to form an exclusive partnership like Rover.com did with Petco, which has invested in its Series B to Series E rounds.

A third option is for start-ups to do a structural capital hack. For example, Roadie sought out UPS Capital to provide it with an insurance solution and brought in UPS as a strategic investor. But, ultimately, if a start-up discovers it is lacking the structural capital to profitably originate, extract, and capture value, it might be better off forgetting about capital hacking and instead find an exit through a strategic buyer. This happened in January 2016 when GM acquired the shuttered assets of Sidecar, which wasn't able to effectively compete against Uber and Lyft and was unsuccessful in its strategic pivot in August 2015 to the goods delivery vertical.

If venture capital flows remain scarce, this could jeopardize the sharing/on-demand economy. Since I published my research report in September 2015, 8 of the top 75 sharing/on-demand economy companies have shut down. And as of my August 2016 count, 19 companies have not raised any capital since the end of 2014. But as I realized from my discussion with Ryan, it does not have to be a doomsday scenario. Established companies could provide start-ups with access to the capital they need in order to survive—not just financial capital but, more importantly, supplier capital, customer capital, or structural capital. And this is creating an opportunity for companies looking to embrace ubernomics.

The Ubernomics Strategy Canvas

In *What Matters Now,* Gary Hamel states, "Today's most successful business models rely on value-creating networks and forms of social production that transcend organizational boundaries."[80] To be honest, I didn't grasp the full depth of Hamel's thinking when I first read his book in February 2012. But now, looking back, I realize how visionary his thinking was, as it speaks to the potential for companies to unlock the hidden value of the underutilized physical and human capital in their stakeholder ecosystem through collaboration.

Forming a strategic partnership with a marketplace start-up makes sense for a company looking for a cost-effective and timely solution to optimizing its physical and human capital or providing a sharing/on-demand economy experience for its customers. However, as Frank Eliason advises in *@Your Service,* "In the new world of @Your Service, everyone who is in contact with your Customer impacts your brand."[81] So companies that are looking for a technology solution to deliver a good or service that involves an interaction with their customers might be better off exploring other options. And even if there is no customer interaction, companies still face the risk of potential disintermediation if they allow a start-up to manage the customer relationship for them.

Another option for a company that is looking to join the sharing/on-demand economy without the operational and brand risk of partnering with a start-up, or the capital and technology risk of trying to build its own platform, is to partner with a SaaS platform provider. This option did not occur to me until November 2014, when I came across an article on LinkedIn titled "The Arrival of Dispatch," written by Avi Goldberg, cofounder and CEO of Dispatch. In the article, Goldberg relates how he was inspired for the idea of Dispatch, a SaaS platform to "power the on-demand economy" by uberfying legacy businesses, after the frustrating experience of trying to get his washing machine fixed. By providing companies with a workforce management platform that is customizable and modular, and that seamlessly integrates into their existing infrastructure, Dispatch is helping companies bridge the communication gap between three key stakeholders in the "last mile" of the service delivery process—company, technician, and homeowner—with the inspiring vision of being a "salesforce for the last mile."[82]

Another company that provides a white-label solution for companies looking to build a sharing economy marketplace is Near Me. As I listened to its founder, Adam Broadway, discuss how Near Me is working with Intel to create a marketplace for its 20 million developers, I couldn't help but think about the untapped potential for other companies to build a marketplace around their respective communities.

The ultimate goal for a company should be to build a proprietary sharing/on-demand economy platform marketplace—one that complements or enhances its core product or service offering. Companies can unlock hidden value either by working with sharing/on-demand economy start-ups (i.e., partnering with, strategically investing in, acquiring) or by collaborating with one or more of their corporate stakeholders (e.g., the firm itself, customers, suppliers, partners, competitors) to build their own marketplace. I visualize this as the ubernomics strategy canvas, which is represented by the 10 marketplace verticals on the y-axis and the eight collaboration options on the x-axis.

For example, GM is focusing on the private drivers vertical through its acquisition of Sidecar and strategic investment in Lyft. Hyatt strategically invested in onefinestay (a personal asset–sharing start-up) and is now collaborating with its competitors (owners of independent boutique hotels) to build a corporate asset–sharing marketplace. LinkedIn is collaborating with its customers (professionals who use its platform) to build a professional services market, while Amazon is collaborating with its customers (Amazon shoppers) to build a marketplace in trades services with Handmade at Amazon and in goods delivery with Amazon Flex.

Now imagine a giant wall covered with 156 individual canvases, representing each of the GIC (general industrial classification) subindustries—with not just 80 white spaces but more than 12,000 white spaces. Over the next decade, we will see significant disruption of industries as companies start to fill in this blank canvas by entering new white spaces and unlocking hidden value through ubernomics—the next generation of business strategy.

Afterword: Returning Home to My Tribe

This is the final stage of the Hero's journey in which he returns home to his Ordinary World a changed man. He will have grown as a person, learned many things... His return may bring fresh hope to those he left behind, a direct solution to their problems or perhaps a new perspective for everyone to consider... Ultimately the Hero will return to where he started but things will clearly never be the same again.

—Joseph Campbell, *The Hero's Journey*

In June 2016, I flew to Toronto to join a former colleague, Heather Hatch, in celebrating the launch of her firm, Kilgharrah Asset Management. Although once upon a time in my career I used to fly to Toronto every month for business, I hadn't been there for nearly a decade. I fantasized about staying at Le Germain Hotel, the high-end boutique hotel that used to be my second home, but as I am bootstrapping Brady Capital Research and no longer have the luxury of a corporate expense account, I couldn't stomach paying its room rate of C$500 per night. And given that one of the objectives of my trip was to share my research on ubernomics, I figured I should practice what I preach, so I looked for a place through Airbnb, and it did not disappoint. I found a gorgeous new condo with an amazing view of the CN Tower for a fifth of the cost of a room at Le Germain (which, ironically, was located right across the street). And although the booking had been for a private room in the condo, the owner decided to take my money and head off to Costa Rica, leaving me with the place to myself.

While it couldn't have worked out better in the end, the process leading up to the booking wasn't as smooth as I would have liked—which provided me with some new insights. While Airbnb has the potential to disrupt business travel (like Uber, which accounted for 43% of ground transportation business travel expenses in 1Q16, according to Certify), my experience made me realize that it needs to overcome several demand compatibility challenges, as the service it provides is not frictionless.

The initial obstacle I encountered was waiting three hours after messaging the first host I contacted (Airbnb gives hosts 24 hours to respond), only to learn that the place was actually not available. It was a stressful process. To circumvent this, when I continued my search, I filtered it to show me only the places available for an "instant book." While doing this narrowed my options, it enabled me to book a place on-demand; however, when I didn't hear back from the host after a few days, I started to question my decision to stay with a complete stranger. But thanks to the democratization of data through platforms like LinkedIn and Twitter, I was able to look up the host's profile and gain some peace of mind. My last challenge was trying to deal with the logistics of picking up the key and then dropping it off for an absentee host.

Once my accommodation was successfully booked, I decided to reach out to my former colleagues and clients in Toronto to see if they would be free to meet up for a coffee—I was eager to catch up with them. I wasn't sure at first how to handle the logistics of this, but thanks to Calendly (a fabulous meeting-scheduling tool that I discovered), I was able to successfully book more than 20 back-to-back meetings over three days. (Coincidentally, only days after I returned from my trip, Microsoft announced its new Starbucks for Outlook add-in, which enables Outlook users to schedule meetings at nearby Starbucks locations. If only it had released this a week earlier, as I had used Starbucks as a home base for nearly all my meetings!) I used LinkedIn's messaging feature to reach out to people to set up meetings—and with Microsoft acquiring LinkedIn, I am hoping LinkedIn will now be able to offer more functionality. It would be useful if it could incorporate apps like Microsoft's Skype and Calendly onto its platform.

Enjoying coffee with familiar faces, and dressed in my Armani suit and heels (as opposed to my normal work attire of lululemon pants and T-shirt), I felt like I had stepped back in time, and I realized the true value of professional relationships with deep roots. Although LinkedIn is an incredibly empowering platform that enables professionals to form bridging capital and effortlessly connect with one another by providing transparency into people's character and credibility, the reality is that it builds only shallow roots. Real trust is earned over time.

The last time I had seen many of my former colleagues and clients was in January 2008, when I had published the highly controversial sell recommendation on Yellow Pages, which had caused the stock to drop over 5% that day. And, as I relate in Chapter 2, although it had taken four years for my thesis to play out, it turned out I was correct in my radical assertion that directory businesses were no longer a source of predicable and sustainable cash flow, as the democratization of content did indeed lead to structural disruption. And so I felt it was fitting that in my ubernomics presentation (which I had prepared in order to more easily share my theory with those I was catching up with and who were interested), the title of the second slide was "From Yellow Pages to Yellow Cabs," since my thesis is that the shift from scarcity to abundance has now advanced from the democratization of content to the democratization of physical and human capital.

It was interesting going through the presentation and seeing the different responses to my radical new thesis. A few people were mainly interested in catching up with me on a personal basis (it was fun to show pictures of my three- and five-and-a-half-year-old boys), but most were interested to hear about my new perspective on value creation through ubernomics, and over a third saw the exciting potential for this next generation of business strategy to create value.

It felt great to have returned home to my tribe.

Acknowledgements: Oracles I Met along My Journey

In addition to the rebel-with-a-cause trickster founders and the mentors, I am grateful for the wise words of the visionary oracles who I met along my journey. The wisdom they were imparting didn't always register with me immediately, but it connected when I was ready to hear it. Let me highlight here some of the people whose wisdom had the biggest influence on me.

Shortly before leaving the corporate world, when I was still working as a sell-side equity analyst, I read Jeff Jarvis's brilliant book *What Would Google Do?*, in which he cautions, "Beware the cash cow in the coal mine."[83]This was something that really resonated with me, as in my former life of analyzing high-dividend-yielding companies, I advised clients to beware of "wolves in cash-cows' clothing." (A warning that holds just as true today for Yellow Cab as it did back then for Yellow Pages.) What I value most about the book, though, is that it introduced me to the revolutionary concept of platforms. As Jarvis explained it, "Networks are built atop platforms… A platform enables. It helps others build value. Any company can be a platform… Platforms help users create products, businesses, communities, and networks of their own." And this book demonstrates that it is through creating platforms that companies are able to achieve economics of abundance.

I encountered Jarvis again in October 2011 (early in my post-corporate era) when I read *Public Parts*. Among the questions that Jarvis probes in this book is one at the heart of abundance economics: "How can a company improve and profit by opening up its information and its processes to transform relationships, collaborate, and profit?"[84] But as I was early in my intellectual journey, I had not yet developed the mental framework to comprehend the true genius of his thinking. For at that point, I was still wandering around the base of the Social Economy Pyramid, exploring how social media was democratizing influence, leading to the new social value driver of advocacy.

Another oracle who had a significant influence on me is Gary Vaynerchuk. In fact, I still joke with my husband, Greg, how it seemed like Vaynerchuk was part of our two-and-a-half-month road trip across the United States in early 2010. Not only did we listen to the audiotape of his first book, *Crush It!*, as we cruised down the Pacific Coast Highway, but we saw him speak at not one, not two, but three conferences. The following year, Vaynerchuk published his second book, *The Thank You Economy*, and I was encouraged when in it I read these prescient words: "In the future, the companies with tremendous 'relationship capital' will be the ones to succeed," as it further strengthened my conviction in my thesis of advocacy as a new social value driver.[85] In retrospect, I wish I had paid as much attention to Vaynerchuk's investment portfolio as I did to his books and speeches. As he says in his LinkedIn profile, he was a "prolific angel investor early on," investing in companies like Facebook, Twitter, Tumblr, Uber, and Birchbox. I remember hearing him talk six years ago at SXSW about how he aspired to buy the New York Jets; if he keeps "crushing it" with his investments, I now have no doubt his dream will one day come true.

Another book that made an impression on me around the same time was Chip Conley's *Peak*, in which Conley explores the idea that great companies have great causes. Drawing on his own experience as the founder of Joie de Vivre Hospitality, Conley introduces his concept of the Employee Pyramid, which illustrates, as he puts it, the "three kinds of relationships one can have with work: you either have a job, a career, or a calling."[86] Looking back, I think this idea helped plant the seed in my mind to apply Maslow's Hierarchy of Needs to come up with my concept of visualizing the social economy as a pyramid.

Conley continued to be a source of inspiration to me. I read his book *Emotional Equations* in 2012, soon after its publication. In it, Conley shares his thoughts on how to live a meaningful life, and there's one particular quote of his that still speaks to me: "The most important challenge might be in finding the willingness to give up who you think you are in order to find out who you might become. That is the path to authenticity."[87] The following year, in April 2013, I had the privilege of hearing Conley present at the Conscious Capitalism Conference in San Francisco, and it was with some fascination that I read six months later that Brian Chesky, the founder of Airbnb, had brought in Conley as his right-hand man, to be the head of Global Hospitality. However, I failed to fully connect the dots until over a year later, when I experienced the first-time magic of staying at an Airbnb.

At the SXSW conference in March 2012, I heard Doc Searls, coauthor of *The Cluetrain Manifesto,* deliver a passionate presentation titled "Are Free Customers Better Than Captive Ones?" Interestingly, the book was published in June 1999, at the height of the dot-com boom, yet it includes a prescient statement, which I was reminded of when I looked back on my notes: "It is said that every economic era is characterized by a set of abundances and a set of scarcities. Firms that can take advantage of the abundances as well as the scarcities are bound to succeed."[88] I wish I had picked up on this clue when I initially read the book, but that wouldn't happen for another two years, until I started to research the sharing/on-demand economy and realized that these companies were defying traditional economic principles of scarcity.

Another publication that proved invaluable to me was Dion Hinchcliffe and Peter Kim's *Social Business by Design.* When I read it in June 2012, I had just published my research report *LinkedIn: Disrupting by the "Power of We,"* and so I could really relate to the authors' statement that "social business is one of the biggest shifts in the structure and process of our organizations in business history."[89] At this point, I had ascended to the middle of the Social Economy Pyramid, as I had discovered that the democratization of data through companies like LinkedIn was leading to the emergence of connection, the second social value driver. I wasn't quite ready yet to heed the authors' other words of wisdom that "peer production is the most efficient and richest source of value creation," which spoke to the emergence of collaboration, the third social value driver—which I wouldn't discover for yet another two years.

Also of great significance to me was *Spend Shift* by John Gerzema and Michael D'Antonio, who wisely observe: "Whereas things defined the credit age, meaning now guides the debit age."[90] The shift from valuing *things* to valuing *meaning* resonated with me, as it confirmed my belief that companies with a greater purpose would thrive in the new social era, as their stakeholders (customers, employees, partners) would form a deep psychological attachment to what companies stand for. Once again my note-taking served me well, for when I reviewed the information I noted from their book, I saw that the authors had offered two important clues that hinted at the structural shifts in societal forces that were giving rise to the sharing economy: "Increasingly we'll buy people and experiences instead of products and services... The companies that succeed will understand the American ideal for a 'liquid life.' They will help their customers de-leverage by offering simplicity, efficiency, and flexibility to adapt to ever-changing conditions."[91]

As you know from reading this book, I only discovered for myself the secret path leading to the peak of the Social Economy Pyramid in May 2014 when I experienced the sharing economy firsthand, using Airbnb and Uber. I find it fascinating to explore this emerging land of tech-enabled abundance. As Peter H. Diamandis and Steven Kotler assert in their book *Abundance*, "Technology is a resource-liberating mechanism. It can make the once scarce the now abundant."[92] But as a member of Generation X who entered the business world in the early 1990s, I can understand how this new paradigm can be confounding to many people. Clay Shirky describes this well in *Cognitive Surplus*: "Because abundance can remove the trade-offs we're used to, it can be disorienting to the people who've grown up with scarcity. When a resource is scarce, the people who manage it often regard it as valuable in itself, without stopping to consider how much of the value is tied to its scarcity."[93]

And last, companies would be wise to heed the warnings of Jeremy Rifkin, who cautions in *The Zero Marginal Cost Society*, "If the people formerly known as consumers begin consuming 10% less and peering 10% more, the effect on margins of traditional corporations is going to be disproportionately greater... Which means certain industries have to rewire themselves, or prepare to sink into the quicksand of the past."[94]

About the Author

Barbara Gray is a former top-ranked sell-side equity analyst and the founder of Brady Capital Research Inc., a leading-edge research and strategy consulting firm. Barbara ranked as an All-Star Analyst for four consecutive years, achieving top-three standing each year in the business trust sector in the Brendan Wood annual institutional survey. She has more than fifteen years of sell-side equity research experience in Canada and the United States covering a wide range of sectors. Barbara has a Bachelor of Commerce (Finance) from the University of British Columbia (1993) and earned her Chartered Financial Analyst (CFA) designation in 1997.

Notes

Introduction: An Analyst's Journey

[1]Disclosure: As of the time of writing, I have a long position in the following companies mentioned in this book: lululemon athletica inc. (LULU-NASDAQ; LLL-TSX), LinkedIn Corporation (LNKD-NYSE), and Zillow Group Inc. (ZG-NASDAQ). I have a short position in Medallion Financial Corporation ($MFIN-NYSE).

1. The Next Generation of E-Commerce

[2]All mission statements and core values quoted in this book were taken from the respective corporate websites.

[3]Seeking Alpha, "Hyatt Hotels (H) Mark S. Hoplamazian on Q3 2015 Results—Earnings Call Transcript," Seeking Alpha, November 3, 2015, http://seekingalpha.com/article/3640646-hyatt-hotels-h-mark-s-hoplamazian-q3-2015-results-earnings-call-transcript.

[4]Chris Anderson, *The Long Tail: Why the Future of Business Is Selling More of Less* (New York: Hyperion, 2006).

[5]W. Chan Kim and Renee Mauborgne, *Blue Ocean Strategy: How to Create Uncontested Market Space and Make the Competition Irrelevant* (Boston: Harvard Business School Press, 2005), 4.

[6]Seth Godin, *The Icarus Deception: How High Will You Fly?* (New York: Portfolio Hardcover, 2012).

[7]Milton Friedman, *Capitalism and Freedom* (Chicago: University of Chicago Press, 1962).

2. A Wake-Up Call for CEOs

[8]Jeff Jarvis, *Public Parts: How Sharing in the Digital Age Improves the Way We Work and Live* (New York: Simon & Schuster, 2011), 179.

[9]Josh Barro, "Under Pressure from Uber, Taxi Medallion Prices Are Plummeting,"

New York Times, November 27, 2014.

[0]Anita McGahan, "How Industries Change," *Harvard Business Review*, October 2004.

[1]Robert J. Shiller, *Finance and the Good Society* (Princeton, NJ: Princeton University Press, 2012).

3. Defying Economic Principles of Scarcity

[2]American Hotel & Lodging Association, *Lodging Industry Trends 2015*, https://www.ahla.com/uploadedFiles/_Common/pdf/Lodging_Industry_Trends_2015.pdf.

[3]Peter Thiel and Blake Masters, *Zero to One: Notes on Startups, or How to Build the Future* (New York: Crown Business, 2014).

[4]Jonah Sachs, *Winning the Story Wars: Why Those Who Tell—and Live—the Best Stories Will Rule the Future* (Boston: Harvard Business Review Press, 2012), 171.

[5]Scott Goodson, *Uprising: How to Build a Brand—and Change the World—by Sparking Cultural Movements* (New York: McGraw-Hill, 2012).

4. The Value of Advocacy

[6]Gary Klein, *Seeing What Others Don't: The Remarkable Ways We Can Gain Insights* (New York: Public Affairs, 2013).

[7]Howard Schultz and Dori Jones Yang, *Pour Your Heart into It: How Starbucks Built a Company One Cup at a Time* (New York: Hyperion, 1997).

[8]Simon Sinek, *Start with Why: How Great Leaders Inspire Everyone to Take Action* (New York: Portfolio, 2011); emphasis in original.

[9]John Mackey, "Conscious Capitalism: Creating a New Paradigm for Business," *John Mackey's Blog*, November 9, 2006, http://www.wholefoodsmarket.com/blog/john-mackeys-blog/conscious-capitalism-creating-new-paradigm-for-business.

[0]Fred Lager, *The Inside Scoop: How Two Real Guys Built a Business with a Social Conscience and a Sense of Humor* (New York: Crown, 1994).

[1]Dev Patnaik, *Wired to Care: How Companies Prosper When They Create Widespread Empathy* (Upper Saddle River, NJ: Pearson Education, 2009).

[2]C. William Pollard, *The Soul of the Firm* (Grand Rapids: Zondervan, 1996), 45.

[3]lululemon athletica inc., *Form 10-K Annual Report for the Fiscal Year Ended January 30, 2011*, https://www.sec.gov/Archives/edgar/data/1397187/000095012311026220/o67665e10vk.htm.

5. How Rebels with a Cause Build Advocacy

[24]Norman Wolfe, *The Living Organization: Transforming Business to Create Extraordinary Results* (Irvine, CA: Quantum Leaders, 2011).

[25]Schultz and Yang, *Pour Your Heart into It.*

[26]lululemon, "2011 lululemon annual report," YouTube video, 6:40, posted June 7, 2012, https://www.youtube.com/watch?v=kY9REMev-Wo.

[27]Frank Eliason, *@Your Service: How to Attract New Customers, Increase Sales, and Grow Your Business Using Simple Customer Service Techniques* (Hoboken, NJ: Wiley, 2012).

[28]Grant McCracken, *Culturematic: How Reality TV, John Cheever, a Pie Lab, Julia Child, Fantasy Football... Will Help You Create and Execute Breakthrough Ideas* (Boston: Harvard Business Review Press, 2012).

[29]Business Wire, "lululemon Founder and Largest Shareholder Chip Wilson Sends Open Letter to lululemon Shareholders," news release, June 1, 2016, http://www.businesswire.com/news/home/20160601006649/en/lululemon-Founder-Largest-Shareholder-Chip-Wilson-Sends.

6. Shift to Social Consciousness Elevates Advocacy

[30]Pip Coburn, *The Change Function: Why Some Technologies Take Off and Others Crash and Burn* (New York: Portfolio, 2006), 194.

[31]Geoff Lewis, "Advocacy Investing—Catnip for Wealthy Clients?" *RegisteredRep*, June 1, 2006.

[32]Nassim Nicholas Taleb, *Antifragile: Things That Gain from Disorder* (New York: Random House, 2012), 9.

[33]Heidi Grant Halvorson and E. Tory Higgins, *Focus: Use Different Ways of Seeing the World for Success and Influence* (New York: Hudson Street Press, 2013).

[34]UBS, *UBS Investor Watch Report Reveals Millennials Are as Financially Conservative as Generation Born During Great Depression*, January 27, 2014, https://www.ubs.com/us/en/wealth/news/wealth-management-americas-news.html/en/2014/01/27/ubs-investor-watch-report-reveals-millennials.html.

[35]Marc Gunther, *Faith and Fortune: How Compassionate Capitalism Is Transforming American Business* (New York: Crown Business, 2005), 33.

[36]Thor Muller and Lane Becker, *Get Lucky: How to Put Planned Serendipity to Work for You and Your Business* (San Francisco: Jossey-Bass, 2012).

[37]Diana Rivenburgh, *The New Corporate Facts of Life: Rethink Your Business to*

Transform Today's Challenges into Tomorrow's Profits (New York: AMACOM, 2013), 46.

7. The Value of Connection

[38] Ori Brafman and Rod A. Beckstrom, *The Starfish and the Spider: The Unstoppable Power of Leaderless Organizations* (New York: Portfolio, 2008).

[39] Jeff Jarvis, *Gutenberg the Geek* (Seattle: Amazon Digital Services, 2012).

[40] Mark S. Granovetter, "The Strength of Weak Ties," *American Journal of Sociology* 68, 6 (1973).

[41] Robert D. Putnam, *Bowling Alone: The Collapse and Revival of American Community* (New York: Simon & Schuster, 2000), 19.

[42] John Hagel III, John Seely Brown, and Lang Davison, *The Power of Pull: How Small Moves, Smartly Made, Can Set Big Things in Motion* (New York: Basic Books, 2010).

[43] Stephen M.R. Covey and Greg Link, *Smart Trust: Creating Prosperity, Energy, and Joy in a Low-Trust World* (New York: Free Press, 2012).

[44] Frans Johansson, *The Medici Effect: Breakthrough Insights at the Intersection of Ideas, Concepts, and Cultures* (Boston: Harvard Business Review Press, 2004).

[45] IBM, *2012 IBM CEO Study*, http://www-935.ibm.com/services/us/en/c-suite/ceostudy2012/.

[46] Joey Reiman, *The Story of Purpose: The Path to Creating a Brighter Brand, a Greater Company, and a Lasting Legacy* (Hoboken, NJ: Wiley, 2012).

[47] Chip Conley, *Emotional Equations: Simple Truths for Creating Happiness + Success in Business + Life* (New York: Atria, 2012).

[48] Reiman, *The Story of Purpose*; emphasis in original.

[49] Dr. Seuss, *There's a Wocket in My Pocket!* (New York: Random House, 1974); emphasis in original.

[50] Zillow Inc., *Form 10-K Annual Report for the Fiscal Year Ended December 31, 2013*, https://www.sec.gov/Archives/edgar/data/1334814/000119312514056800/d625583d10k.htm.

[51] Don Tapscott and David Ticoll, *The Naked Corporation: How the Age of Transparency Will Revolutionize Business* (New York: Free Press, 2003), 27.

[52] Umair Haque, *Betterness: Economics for Humans* (Boston: Harvard Business Review Press, 2011).

[53] John Gerzema and Michael D'Antonio, *The Athena Doctrine: How Women (and the Men Who Think like Them) Will Rule the Future* (San Francisco: Jossey-Bass, 2013).

[54]Tom Asacker, *Opportunity Screams: Unlocking the Hearts and Minds of Today's Idea Economy* (Ithaca, NY: Paramount Marketing, 2010).

8. The Value of Collaboration

[55]Clayton M. Christensen and Michael E. Raynor, *The Innovator's Solution: Creating and Sustaining Successful Growth* (Boston: Harvard Business Review Press, 2003).

[56]Thomas A. Stewart, *Intellectual Capital: The New Wealth of Organizations* (New York: Doubleday Currency, 1998).

[57]Heather McGowan, "Jobs Are Over: The Future Is Income Generation: Part 2," LinkedIn, June 22, 2014, https://www.linkedin.com/pulse/20140623011119-22726740-jobs-are-over-the-future-is-income-generation-part-2.

[58]J.C. Larreche, *The Momentum Effect: How to Ignite Exceptional Growth* (Harlow, PA: Wharton School Publishing, 2008).

[59]Stewart, *Intellectual Capital.*

[60]Larreche, *The Momentum Effect.*

[61]Anderson, *The Long Tail.*

[62]Kim and Mauborgne, *Blue Ocean Strategy.*

[63]Clayton M. Christensen, Mark W. Johnson, and Darrell K. Rigby, "Foundations for Growth: How to Identify and Build Disruptive New Businesses," *MIT Sloan Management Review* 43, 3 (2002).

[64]Clayton M. Christensen, Michael E. Raynor, and Rory McDonald, "What Is Disruptive Innovation?" *Harvard Business Review*, December 2015.

9. The Sharing/On-Demand Economy

[65]Chris Brogan and Julien Smith, *The Impact Equation: Are You Making Things Happen or Just Making Noise?* (London: Portfolio, 2012).

[66]Jeff Bercovici, "Yet Another On-Demand Startup Exits the Gig Economy," *Inc.*, August 6, 2015.

[67]Dan Lyons, *Disrupted: My Misadventure in the Start-Up Bubble* (New York: Hachette Books, 2016).

[68]Douglas Holt and Douglas Cameron, *Cultural Strategy: Using Innovative Ideologies to Build Breakthrough Brands* (New York: Oxford University Press, 2010).

[69]Dave Logan, John King, and Halee Fischer-Wright, *Tribal Leadership: Leveraging Natural Groups to Build a Breakthrough Organization* (New York: HarperBusiness,

2008).

[70]Mitch Joel, *Ctrl Alt Delete: Reboot Your Business; Reboot Your Life; Your Future Depends on It* (New York: Business Plus, 2013).

[71]Saul J. Berman, *Not for Free: Revenue Strategies for a New World* (Boston: Harvard Business Review Press, 2011).

10. The New White Space

[72]Hyatt Hotels Corporation, "Hyatt Launches New Brand: The Unbound Collection by Hyatt," press release, March 2, 2016, http://investors.hyatt.com/investor-relations/news-and-events/financial-news/financial-news-details/2016/Hyatt-Launches-New-Brand-The-Unbound-Collection-by-Hyatt.

[73]Seeking Alpha, "Hyatt Hotels."

[74]Seeking Alpha, "Hilton Worldwide's (HLT) CEO Chris Nassetta on Q4 2015 Results—Earnings Call Transcript,"Seeking Alpha, February 26, 2016, http://seekingalpha.com/article/3936456-hilton-worldwide-holdings-hlt-christopher-j-nassetta-q4-2015-results-earnings-call-transcript.

[75]Nick Carey and Mari Saito, "With Deliv Investment, UPS Hopes to Study Same-Day Delivery Market," Reuters, February 24, 2016.

[76]Scott D. Anthony, *The Little Black Book of Innovation: How It Works, How to Do It* (Boston: Harvard Business Review Press, 2011), 28.

[77]Harley Manning and Kerry Bodine, *Outside In: The Power of Putting Customers at the Center of Your Business* (Boston: New Harvest, 2012).

[78]Stephanie Gleason and Ted Mann, "GE Says Quirky Has Hurt Its Reputation," *Wall Street Journal*, December 3, 2015, http://www.wsj.com/articles/ge-says-quirky-has-hurt-its-reputation-1449179311.

[79]"Whole Foods Market® and Instacart Partner to Offer One-Hour Delivery across 15 Major U.S. Cities," Whole Foods Market Newsroom, September 8, 2014, http://media.wholefoodsmarket.com/news/.

[80]Gary Hamel, *What Matters Now: How to Win in a World of Relentless Change, Ferocious Competition, and Unstoppable Innovation* (San Francisco: Jossey-Bass, 2012).

[81]Eliason, *@Your Service.*

[82]Avi Goldberg, "The Arrival of Dispatch," LinkedIn, November 18, 2014, https://www.linkedin.com/pulse/20141118143050-4098006-the-arrival-of-dispatch.

Acknowledgements: Oracles I Met along My Journey

[83]Jeff Jarvis, *What Would Google Do? Reverse Engineering the Fastest Growing Company in the History of the World* (New York: Harper Collins, 2009).
[84]Jarvis, *Public Parts.*
[85]Gary Vaynerchuk, *The Thank You Economy* (New York: HarperBusiness, 2011).
[86]Chip Conley, *Peak: How Great Companies Get Their Mojo from Maslow* (San Francisco: Jossey-Bass, 2007), 84.
[87]Conley, *Emotional Equations.*
[88]Rick Levine et al., *The Cluetrain Manifesto: The End of Business as Usual* (Philadelphia: Basic Books, 2001), 279.
[89]Dion Hinchcliffe and Peter Kim, *Social Business by Design: Transformative Social Media Strategies for the Connected Company* (San Francisco: Jossey-Bass, 2012), 20.
[90]John Gerzema and Michael D'Antonio, *Spend Shift: How the Post-Crisis Values Revolution Is Changing the Way We Buy, Sell, and Live* (San Francisco: Jossey-Bass, 2010), 209.
[91]Gerzema and D'Antonio, *Spend Shift*, 209.
[92]Peter H. Diamandis and Steven Kotler, *Abundance: The Future Is Better Than You Think* (New York: Free Press, 2014), 4.
[93]Clay Shirky, *Cognitive Surplus: How Technology Makes Consumers into Collaborators* (New York: Penguin Books, 2011).
[94]Jeremy Rifkin, *The Zero Marginal Cost Society: The Internet of Things, the Collaborative Commons, and the Eclipse of Capitalism* (New York: St. Martin's Press, 2014).

Made in the USA
Charleston, SC
27 October 2016